SpringerBriefs in Optimization

Series Editors

Panos M. Pardalos
János D. Pintér
Stephen M. Robinson
Tamás Terlaky
My T. Thai

SpringerBriefs in Optimization showcases algorithmic and theoretical techniques, case studies, and applications within the broad-based field of optimization. Manuscripts related to the ever-growing applications of optimization in applied mathematics, engineering, medicine, economics, and other applied sciences are encouraged.

For further volumes:
http://www.springer.com/series/8918

Urmila Diwekar • Amy David

BONUS Algorithm for Large Scale Stochastic Nonlinear Programming Problems

 Springer

Urmila Diwekar
Clarendon Hills
Illinois
USA

Amy David
Krennert School of Business
Purdue University
West Lafayette
Indiana
USA

ISSN 2190-8354   ISSN 2191-575X (electronic)
SpringerBriefs in Optimization
ISBN 978-1-4939-2281-9   ISBN 978-1-4939-2282-6 (eBook)
DOI 10.1007/978-1-4939-2282-6

Library of Congress Control Number: 2014955715

Springer New York Heidelberg Dordrecht London

Printed on acid-free paper

Springer is part of Springer Science+Business Media (www.springer.com)

To my husband Dr. Sanjay Joag,
who changed his career to support mine,
for his constant love and support.
                              —Urmila

Preface

Stochastic programming problems are very difficult to solve as they involve optimization as well as uncertainty analysis. Algorithms for solving large-scale nonlinear stochastic programming problems are very few in number, as are the engineering applications of these problems. This book introduces two algorithms for large-scale stochastic nonlinear problems for both open equation systems and black box models. These algorithms are the Better Optimization of Nonlinear Uncertain Systems (BONUS) algorithm and the L-shaped BONUS algorithm. Real-world applications of these algorithms in the areas of energy and environmental engineering are also detailed. Many have contributed to this book. Researchers who worked with Dr. Diwekar including Dr. Adrian Lee, Dr. Kemal Sahin, Dr. Juan Salazar, and Dr. Yogendra Shastri, as well as collaborators such as Dr. Emil Constantinescu, Dr. Victor Zavala, and Dr. Stephen Zitney have provided the material for this book with their research. Thanks also to our group members Dr. Pahola Benavides, Dr. Berhane Gabreslassie, Dr. Rajib Mukherjee, Shivam Tyagi, and Kirti Yenki who went through the first draft of the book and meticulously pointed out mistakes. Hope you enjoy this work.

Urmila Diwekar and Amy David

Contents

List of Figures

List of Tables

Chapter 1
Introduction

A general optimization problem can be stated as follows.

$$\text{Optimize}\quad Z = z(x) \tag{1.1}$$
$$x$$

subject to

$$h(x) = 0 \tag{1.2}$$
$$g(x) \le 0 \tag{1.3}$$

The goal of an optimization problem is to determine the decision variables x that optimize the objective function Z(Eq. 1.1), while ensuring that the model operates within established limits enforced by the equality constraints h (Eq. 1.2) and inequality constraints g (Eq. 1.3).

Figure 1.1 illustrates schematically the iterative procedure employed in a numerical optimization technique. As seen in the figure, the optimizer invokes the model with a set of values of decision variables x. The model simulates the phenomena and calculates the objective function and constraints. This information is utilized by the optimizer to calculate a new set of decision variables. This iterative sequence is continued until the optimization criteria pertaining to the optimization algorithm is satisfied. If the objective function and constraints are linear and the decision variables involved are scalar and continuous, then it is a linear programming (LP) problem. However, if the objective function and/or constraints are nonlinear then it is a nonlinear programming (NLP) problem. An NLP problem involving integers is a mixed integer nonlinear programming (MINLP) problem.

1.1 Stochastic Optimization Problems

Stochastic optimization gives us the ability to optimize systems in the face of uncertainties. A stochastic optimization or a stochastic programming (SP) problem requires that the objective function and constraints be expressed in terms of some

© Urmila Diwekar, Amy David 2015   1
U. Diwekar, A. David, *BONUS Algorithm for Large Scale Stochastic Nonlinear
Programming Problems,* SpringerBriefs in Optimization, DOI 10.1007/978-1-4939-2282-6_1

Fig. 1.1 Pictorial representation of the numerical optimization framework [7]

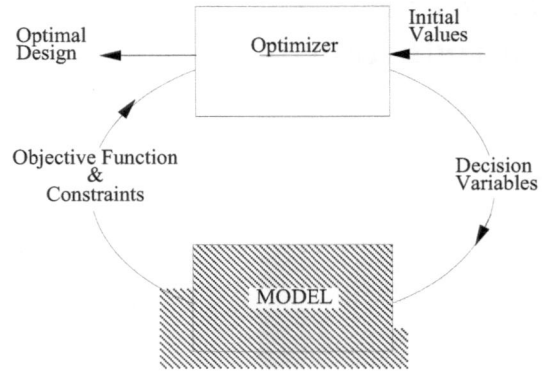

probabilistic representation (e.g., expected value, variance, fractiles, most likely values). For example, in chance constrained programming, the objective function is expressed in terms of expected value, and the constraints are expressed in terms of fractiles (probability of constraint violation), and in Taguchi's offline quality control method ([55], Diwekar and Rubin 1991), the objective is to minimize variance. These problems can be further classified as stochastic linear programming (SLP), stochastic nonlinear programming (SNLP), and stochastic mixed integer linear and nonlinear programming problems. The latter problems are the focus of this book.

A generalized stochastic optimization problem, where the decision variables and uncertain parameters are separated, can then be viewed as:

$$\text{Optimize} \; J \; = \; P_1(j(x,u)) \tag{1.4}$$
$$x$$

subject to

$$P_2(h(x,u)) \; = \; 0 \tag{1.5}$$
$$P_3(g(x,u) \geq 0) \; \geq \; \alpha \tag{1.6}$$

where u is the vector of uncertain parameters and P represents the cumulative distribution functional such as the expected value, mode, variance, or fractiles. Figures 1.2 and 1.3 show the expected value (mean), mode, variance, and fractiles for a probabilistic distribution function.

Unlike the deterministic optimization problem, in stochastic optimization one has to consider the probabilistic functional of the objective function and constraints. The generalized treatment of such problems is to use probabilistic or stochastic models instead of the deterministic model inside the optimization loop. Figure 1.4 represents the generalized solution procedure, where the deterministic model is replaced by an iterative stochastic model with a sampling loop representing the discretized uncertainty space. The uncertainty space is represented in terms of the moments such as the mean, or the standard deviation of the output over the sample space of N_{samp} as given by the following equations (Eqs. 1.7 and 1.8).

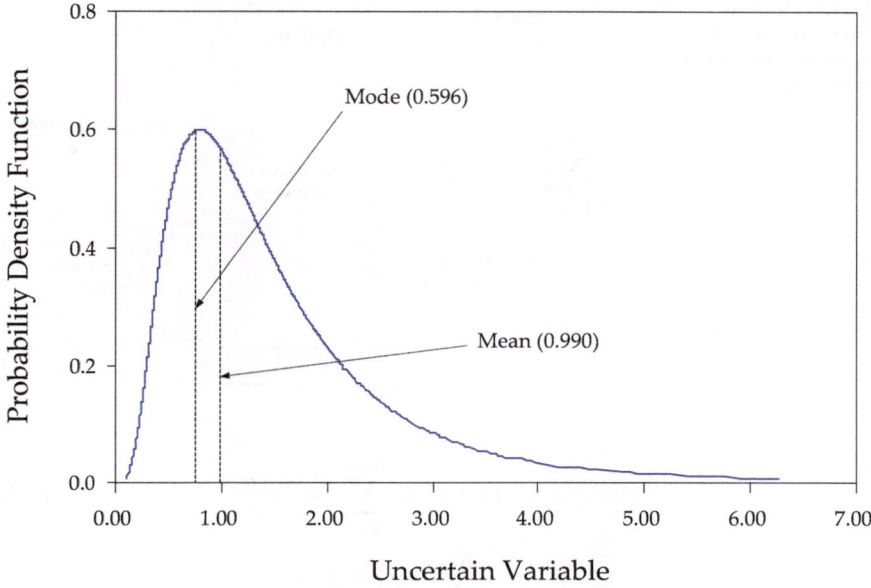

Fig. 1.2 Different probabilistic performance measures (PDF) [7]

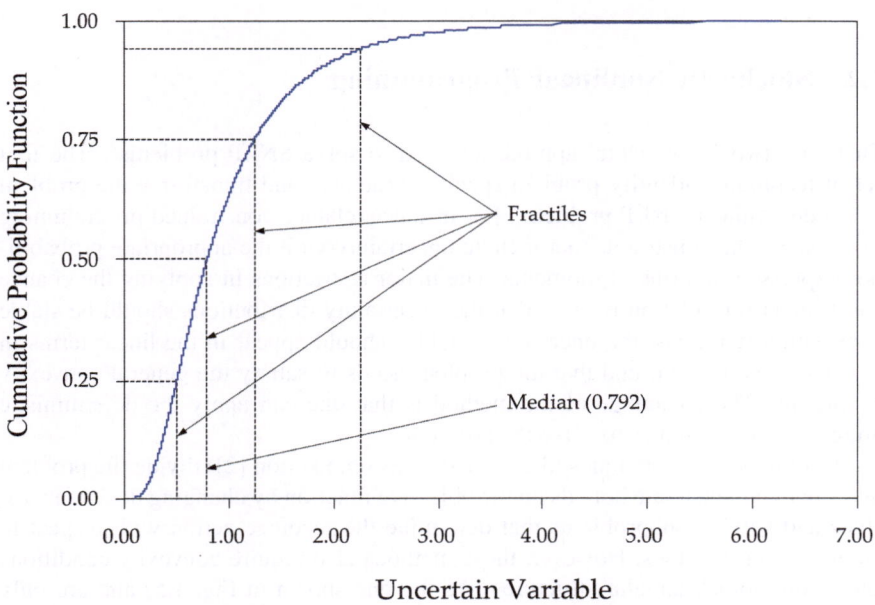

Fig. 1.3 Different probabilistic performance measures (CDF) [7]

Fig. 1.4 Pictorial representation of the stochastic programming framework

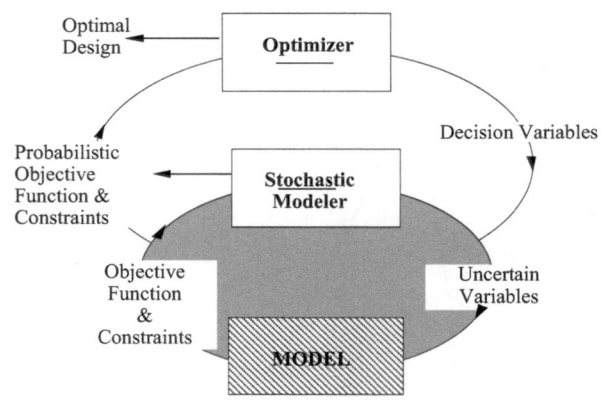

$$E(z(x,u)) = \sum_{k=1}^{N_{samp}} \frac{z(x,u_k)}{N_{samp}} \tag{1.7}$$

$$\sigma^2(z(x,u)) = \sum_{k=1}^{N_{samp}} \frac{(z(x,u_k) - \bar{z})^2}{N_{samp}} \tag{1.8}$$

where $\bar{z}$ is the average value of z. E is the expected value and σ^2 is the variance.

1.2   Stochastic Nonlinear Programming

There are two fundamental approaches used to solve SNLP problems . The first set of techniques identify problem specific structures and transforms the problem into a deterministic NLP problem . For instance, chance constrained programming [4] replaces the constraints that include uncertainty with the appropriate probabilities expressed in terms of moments. The major restrictions in applying the chance constrained formulation include that the uncertainty distributions should be stable distribution functions, the uncertain variables should appear in the linear terms in the chance constraint, and that the problem needs to satisfy the general convexity conditions. The advantage of the method is that one can apply the deterministic optimization techniques to solve the problem.

Decomposition techniques like L-shaped decomposition [2]) divide the problem into stages and generate bounds on the objective function by changing decision variables and solving subproblems that determine the recourse action with respect to the uncertain variables. However, these methods also require convexity conditions and/or dual-block angular structures like the one shown in Fig. 1.5, and are only applicable to discrete probability distributions. For example, Lagrangian-based approaches have been applied to nonlinear SP formulations. The Lagrangian dual ascent

$$
\begin{aligned}
\min \quad & c_0^T x_0 \;+\; c_1^T x_1 \;+\; c_2^T x_2 \;+\; \cdots \;+\; c_N^T x_N \\
\text{s.t.} \quad & A x_0 && = \; b_0, \\
& T_1 x_0 \;+\; W_1 x_1 && = \; b_1, \\
& T_2 x_0 \qquad\qquad +\; W_2 x_2 && = \; b_2, \\
& \;\;\vdots \qquad\qquad\qquad\qquad \ddots && \quad\;\; \vdots \\
& T_N x_0 \qquad\qquad\qquad\qquad\qquad +\; W_N x_N && = \; b_N, \\
& x_0 \ge 0, \quad x_1 \ge 0, \quad x_2 \ge 0, \quad \ldots, \quad x_N \ge 0.
\end{aligned}
$$

Fig. 1.5 Example of a dual block angular structure (LP), each diagonal block is a realization of a random variable(scenario or sample)

method has been proposed by Rockafellar and Wets [38] for problems with finite outcomes for the uncertain variables. Another technique is Regularized Decomposition, which adds quadratic terms to the objective for improved convergence [42] of the L-shaped decomposition method. The augmented Lagrangian method adds a quadratic penalty to ensure convexity, yielding more efficient computation (Dempster, 1988). Rockafellar and Wets also developed a similar technique, the progressive hedging algorithm [39] . These methods have limitations in terms of handling uncertain variables. An alternative approach that can be used to capture uncertainty is through a sampling loop that is embedded within the optimization iterations for the decision variables as shown in Fig. 1.4. This step can be computationally expensive as the model has to be rerun for every sample point. Therefore, we consider efficient sampling techniques. These techniques are described in Chap. 2.

Stochastic approximation methods use upper and/or lower bounds for expected function in a two-stage SP problems. For specific structures (e.g., dual block angular structure) where the L-shaped method is applicable, two approaches consider bounding approximations by embedding sampling within another algorithm without complete optimization. These two approaches are the method of Dantzig [5] , which uses importance sampling to reduce variance in each cut based on a large sample, and the stochastic decomposition method proposed by [20] , which uses the lower bound of the recourse function based on expectation. Again these methods exploit specific structures of the problem and require convexity conditions.

In the BONUS algorithm presented in this book, a generalized approach is proposed that can be used for real world large scale systems without any assumptions. The approach takes advantage of traditional NLP methods such as sequential quadratic programming (SQP) or generalized reduced gradient (GRG2) method that are based on derivative estimation. For real world large scale systems, perturbation derivatives are commonly employed. In order to use derivative-based methods, we need smooth probability density functions; Chap. 3 presents the kernel density estimation (KDE) approach for smooth probability density functions.

Optimization under uncertainty involves iteratively improving a probabilistic objective function. Figure 1.6a shows uncertain input variables with an underlying probability density function $pdf_{in}(x^k, u^k)$, shown as the solid triangular distribution.

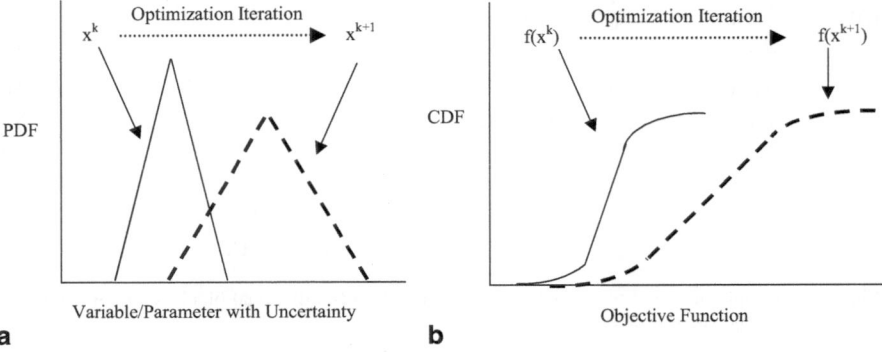

Fig. 1.6 Effect of changes in decision variables

After the model is run, the corresponding output distribution $cdf_{out}(x^k, u^k)$ is generated, shown as the solid line in Fig. 1.6b. As optimization progresses to the next iteration, $k+1$, moments such as mean, variance, and the probability function can change for the uncertain variables, resulting in a new $\widetilde{pdf_{in}}(x^{k+1}, u^{k+1})$, indicated by the dashed line in Fig. 1.6a .

The goal is to identify rapid and efficient techniques that determine an approximation of the properties of the new output distribution, $\widetilde{cdf_{out}}(x^{k+1}, u^{k+1})$, given as the dashed cumulative distribution function in Fig. 1.6b. The advantage of this approach is its bypassing of the model evaluations for successive sampling (the inner loop in Fig. 1.4), which is computationally the most intensive task for optimization under uncertainty. The BONUS algorithm only uses sampling for the first iteration. Details of the algorithm are described in Chap. 4. This is followed by three chapters on application of the BONUS to real world systems. Chapter 8 presents a variant of the BONUS algorithm called L-Shaped BONUS that exploits specific structure problems. Applications of this variant are presented in the Chaps. 9 and 10.

1.3 Summary

A generalized way of solving SNLP is to use sampling-based methods. BONUS exploits the advantages of traditional NLP methods based on derivative information. It uses efficient sampling techniques and uses sampling only for the first iteration. A variant of the the BONUS algorithm, namely, the L-shaped BONUS algorithm uses specific structure of the problem for efficiency improvement. The BONUS algorithm and its variant allows for solution of large scale real world problems.

Notations

$cdf_{out}()$	cumulative probability density function of output
E	expected value function
g	inequality constraint function
h	equality constraint function
J	objective function
N_{samp}	number of samples
$P_i()$	probabilistic function
$pdf_{in}()$	probability density function of input
u	uncertain variable
x	decision variables
Z, z	objective function

Greek letters

σ	standard deviation

Chapter 2
Uncertainty Analysis and Sampling Techniques

The probabilistic or stochastic modeling (Fig. 2.1) iterative loop in the stochastic optimization procedure (Fig. 1.4 in Chap. 1) involves:

1. Specifying the uncertainties in key input parameters in terms of probability distributions
2. Sampling the distribution of the specified parameter in an iterative fashion
3. Propagating the effects of uncertainties through the model and applying statistical techniques to analyze the results

2.1 Specifying Uncertainty Using Probability Distributions

To accommodate the diverse nature of uncertainty, different distributions can be used. Some of the representative distributions are shown in Fig. 2.2. The type of distribution chosen for an uncertain variable reflects the amount of information that is available. For example, the uniform and loguniform distributions represent an equal likelihood of a value lying anywhere within a specified range, on either a linear or logarithmic scale, respectively. Furthermore, a normal (Gaussian) distribution reflects a symmetric but varying probability of a parameter value being above or below the mean value. In contrast, lognormal and some triangular distributions are skewed such that there is a higher probability of values lying on one side of the median than the other. A beta distribution provides a wide range of shapes and is a very flexible means of representing variability over a fixed range. Modified forms of these distributions, uniform* and loguniform*, allow several intervals of the range to be distinguished. Finally, in some special cases, user-specified distributions can be used to represent any arbitrary characterization of uncertainty, including chance distribution (i.e., fixed probabilities of discrete values).

© Urmila Diwekar, Amy David 2015   9
U. Diwekar, A. David, *BONUS Algorithm for Large Scale Stochastic Nonlinear*
Programming Problems, SpringerBriefs in Optimization, DOI 10.1007/978-1-4939-2282-6_2

Fig. 2.1 The stochastic
modeling framework

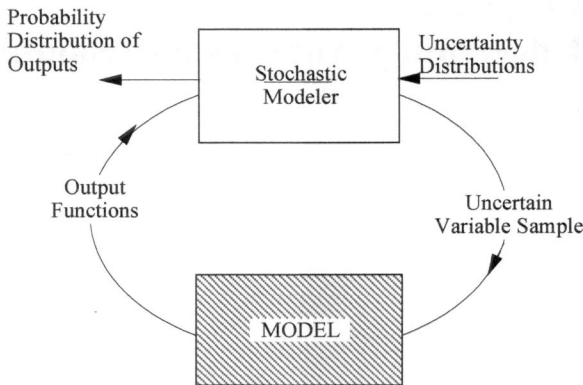

2.2 Sampling Techniques

Sampling is a statistical procedure which involves selecting a limited number of observations, states, or individuals from a population of interest. A sample is assumed to be representative of the whole population to which it belongs. Instead of evaluating all the members of the population, which would be time-consuming and costly, sampling techniques are used to infer some knowledge about the population. Sampling techniques can be divided into two groups: probability sampling and nonprobability

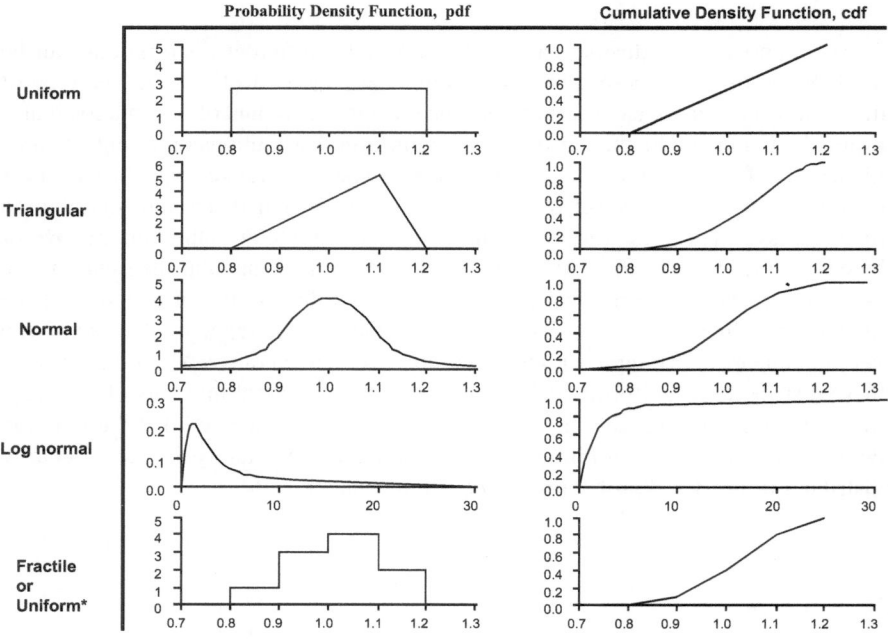

Fig. 2.2 Examples of probabilistic distribution functions for stochastic modeling

sampling. Probabilistic sampling techniques are based on Monte Carlo methods and are most relevant to this chapter. They are described in three subsections below. The description of the sampling techniques below is derived from the sampling chapter by Diwekar and Ulas [10].

2.2.1 Monte Carlo Sampling

One of the simplest and most widely used methods for sampling is the Monte Carlo method. Monte Carlo methods are numerical methods which provide approximate solutions to a variety of physical and mathematical problems by random sampling. The name Monte Carlo, which was suggested by Nicholas Metropolis, takes its name from a city in the Monaco principality which is famous for its casinos, because of the similarity between statistical experiments and the random nature of the games of chance such as roulette.

Monte Carlo methods were originally developed for the Manhattan Project during World War II, to simulate probabilistic problems related to random neutron diffusion in fissile material. Although they were limited by the computational tools of that time, they became widely used in many branches of science after the first electronic computers were built in 1945. The first publication which presents the Monte Carlo algorithm is probably by Metropolis and Ulam [33].

The basic idea behind Monte Carlo simulation has been that input samples should be randomly generated in order to describe a random output. In a crude Monte Carlo approach, a value is drawn at random from the probability distribution for each input, and the corresponding output value is computed. The entire process is repeated n times producing n corresponding output values. These output values constitute a random sample from the probability distribution over the output induced by the probability distributions over the inputs. The simplest distribution that is approximated by the Monte Carlo method is a uniform distribution $U(0, 1)$ with n samples on a k-dimensional unit hypercube. One advantage of this approach is that the precision of the output distribution may be estimated using standard statistical techniques. On average the error of approximation is of the order $O(N^{-1/2})$. One remarkable feature of this sampling technique is that the error bound is not dependent on the dimension k. However, this bound is probabilistic, which means that there is never any guarantee that the expected accuracy will be achieved in a concrete calculation.

The success of a Monte Carlo calculation depends on the choice of an appropriate random sample. The required random numbers and vectors are generated by the computer in a deterministic algorithm. Therefore, these numbers are called pseudorandom numbers or pseudorandom vectors. One of the oldest and best known methods for generating pseudorandom numbers for Monte Carlo sampling is the linear congruential generator (LCG) first introduced by D. H. Lehmer [30]. The general

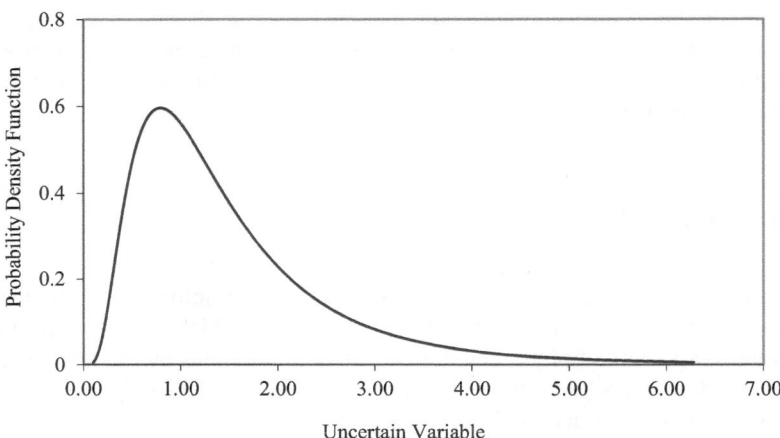

Fig. 2.3 PDF for a lognormal distribution. *PDF probability density function*

formula for a linear congruential generator is the following:

$$I_n = (aI_{n-1} + c)mod\ m \tag{2.1}$$

In this formula, a is the multiplier, c is the increment which is typically set to zero, and m is the modulus. These are preselected constants. The proper choice of these constants is very important for obtaining a sample which performs well in statistical tests. One other preselected constant is the seed, I_0 which is the first number in the output of a linear congruential generator. The random number generator used for Monte Carlo sampling provides a uniform distribution $U(0, 1)$. The specific values of each variable are selected by inverse transformation over the cumulative probability distribution. The following example shows how to generate a sample from pseudorandom numbers.

Example 2.1 We generated four pseudorandom numbers for sampling. These random numbers are $I_n = 0.6, 0.25, 0.925, 0.850$. Find the Monte Carlo samples for the lognormal distribution shown in Fig. 2.3.

Solution From the PDF shown in Fig. 2.3, we created the CDF (Fig. 2.4). We use the y-axis of Fig. 2.4 to place the random numbers on the figure and selected the corresponding x-axis numbers as samples in Table 2.1.

Pseudorandom numbers of different sample sizes on a unit square generated using the linear congruential generator are given in Fig. 2.5. From this figure it can be seen that the pseudorandom number generator produces samples that may be clustered in certain regions of the unit square and does not produce uniform samples. Therefore, in order to reach high accuracy, larger sample sizes are needed, which adversely affects the efficiency of this method. Variance reduction techniques address this problem of increasing efficiency of Monte Carlo methods and are described in the following section.

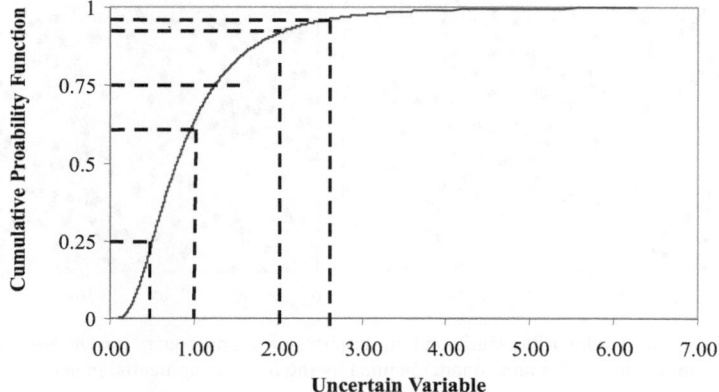

Fig. 2.4 Sample placement on the CDF. *CDF cumulative density function*

Table 2.1 Sample generation

Sample no.	Random number	Sample
1	0.6	1.0
2	0.25	0.5
3	0.925	2.6
4	0.850	2.0

2.3   Variance Reduction Techniques

To increase the efficiency of Monte Carlo simulations and overcome disadvantages such as probabilistic error bounds, variance reduction techniques have been developed [23].

The sampling approaches for variance reduction that are used most frequently in optimization under uncertainty are: importance sampling, Latin Hypercube Sampling (LHS) [22, 32], descriptive sampling, and Hammersley sequence sampling (HSS) [24]. The latter technique belongs to the group of quasi-Monte Carlo methods which were introduced in order to improve the efficiency of Monte Carlo methods by using quasi-random sequences that show better statistical properties and deterministic error bounds. These commonly used sampling techniques are described below with examples.

2.3.1   Importance Sampling

Importance sampling, which may also be called biased sampling, is a variance reduction technique for increasing the efficiency of Monte Carlo algorithms. Monte

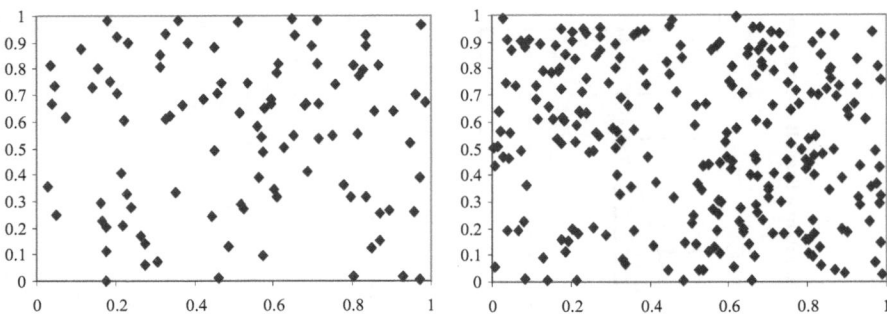

Fig. 2.5 (*Left hand side*) 100 pseudorandom numbers on a unit square, (*right hand side*) 250 pseudorandom numbers on a unit square obtained by the linear congruential generator developed by Wichmann and Hill [62]

Carlo methods are commonly used to integrate a function F over the domain D:

$$I = \int_D F(x)dx \tag{2.2}$$

The Monte Carlo integration for this function can be written as:

$$I_{mcs} = \frac{1}{N} \sum_{i=1}^{N} F(x_i) \tag{2.3}$$

where x_i are random numbers generated from a uniform distribution and N corresponds to number of samples.

If random numbers are drawn from a uniform distribution, information is spread over the interval we are sampling over. However, if a nonuniform (biased) distribution $G(x)$ (which draws more samples from the areas which make a substantial contribution to the integral)is used, the approximation of the integral will be more accurate and the process will be more efficient. This is the basic idea behind importance sampling, where a weighting function is used to approximate the integral as follows.

$$I_{imp} = \frac{1}{n} \sum_{i=1}^{n} \frac{F(x_i)}{G(x_i)} \tag{2.4}$$

Importance sampling is crucial for sampling low-probability events. We will revisit importance sampling when we consider the reweighting scheme in the BONUS algorithm in Chap. 5. The most critical issue for the implementation of importance sampling is the choice of the biased distribution which emphasizes the important regions of the input variables. A simple example for the application of importance sampling for estimation of a simple integral is given below.

Fig. 2.6 The function behavior

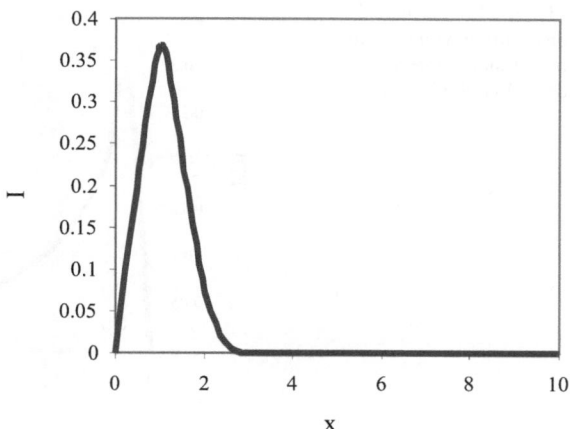

Example 2.2 Integrate the following function using the Monte Carlo method and the method of importance sampling.

$$I = \int_0^{\inf} x^2 \exp(-x^2)dx \qquad (2.5)$$

Solution This function is not possible to integrate analytically but its value is known to be $\sqrt{\pi}/4 = 0.44311328\ldots$. As can be observed from Fig. 2.6, the value of this function decreases rapidly when x is greater than about 3.5. Therefore, there are only a small number of input arguments x where the integral has an appreciable value. If we apply a Monte Carlo integration to estimate this integral, we can uniformly sample the domain of this integral by using a uniform distribution between 0 and 1000 (a large value) and evaluate the integral.

However, we know that this integral only has an appreciable value at a specific interval. Because of that, if we use a uniform sample, most of the points will be from areas that correspond to values where the integral has a very small value. Therefore, we can use a nonuniform distribution function instead, for sampling. If we choose a distribution like the lognormal distribution, the number of samples required to obtain an accurate estimation will be less. For example, let us consider a lognormal distribution with mean $\mu = 1$ and a standard deviation of $\sigma = 1.7$. This is shown in Fig. 2.7. We can see that if we use a lognormal distribution, we will be sampling more from the areas of importance that make a significant contribution to the integral. The estimation of this integral using a uniform sample and a lognormal sample is compared in Table 2.2. As we can see, the integral is accurately estimated using importance sampling after only 100 samples. However, it requires 10,000 samples with the crude Monte Carlo method where a uniform distribution is used.

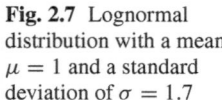

Fig. 2.7 Lognormal distribution with a mean $\mu = 1$ and a standard deviation of $\sigma = 1.7$

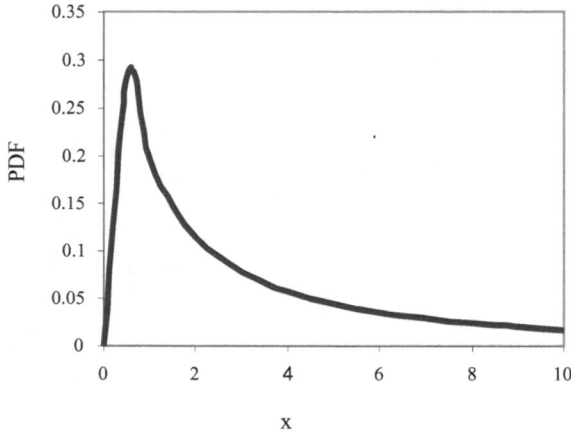

Table 2.2 The estimation of the integral by using uniform random sampling and importance sampling

N	Uniform random sampling	Importance sampling
10	0	0.11054
100	0.00095	0.44363
1000	0.07585	0.44312
10000	0.44131	0.44311

2.3.2   Stratified Sampling

Stratification is the grouping of the members of a population into equal or unequal probability areas (strata) before sampling. The strata must be mutually exclusive, which means that every element in the population must be assigned to only one stratum. Also, no population element is excluded. It is required that the proportion of each stratum in the sample should be the same as in the population.

Latin Hypercube Sampling (LHS) is one form of stratified sampling that can yield more precise estimates of the distribution function [32] and therefore reduce the number of samples required to improve computational efficiency. It is a full stratification of the sampled distribution with a random selection inside each stratum. In LHS, the range of each uncertain parameter X_i is subdivided into nonoverlapping intervals of equal probability. One value from each interval is selected at random with respect to the probability distribution in the interval. The n values thus obtained for X_1 are paired in a random manner (i.e., equally likely combinations) with n values of X_2. These n values are then combined with n values of X_3 to form n-triplets, and so on, until n k-tuples are formed. To clarify how intervals are formed, consider the simple example given below.

Example 2.3   Consider two uncertain variables X_1 and X_2. X_1 has a normal distribution with a mean value of $\mu = 8$ and a standard deviation of $\sigma = 1$. X_2 has a uniform distribution between 5 and 10. Generate an LHS sample for $n = 5$.

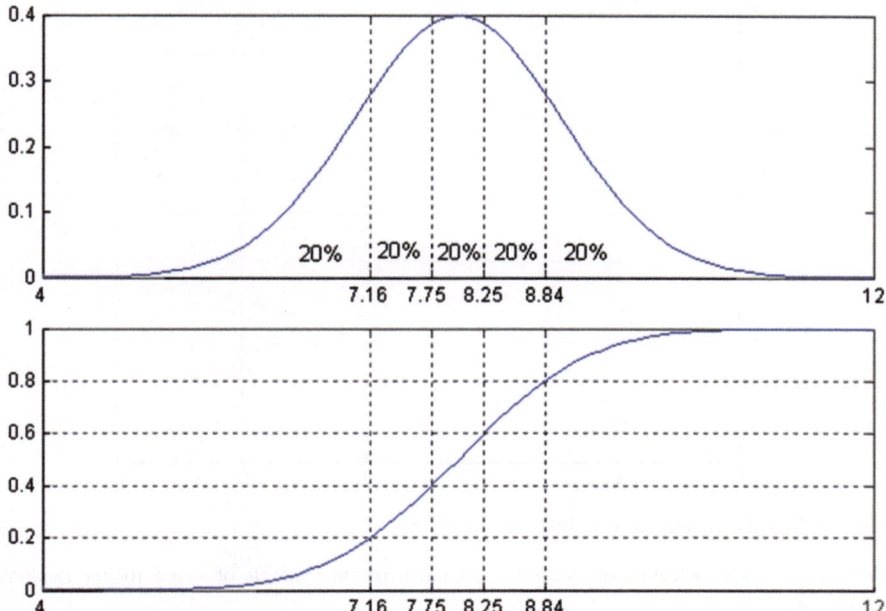

Fig. 2.8 Distribution and stratification for variable X_1

Solution   Figure 2.8 shows the normal distribution PDF and CDF generated using the mean and standard deviation for X_1 and Fig. 2.9 shows the uniform distribution. For LHS, we divide each distribution into equal probability strata. Therefore, we have divided each distribution with five intervals with a 20 % probability each.

The next step to obtain a Latin hypercube sample is to choose specific values of X_1 and X_2 in each of their five respective intervals. This selection is done in a random manner with respect to density in each interval. Next the selected values of X_1 and X_2 are paired randomly to form the 2-dimensional input vectors of size 5. This pairing is done by a random permutation of the first 5 integers with each input variable. For example, we can consider two random permutations of the integers (1, 2, 3, 4, 5):

Permutation 1: (2, 5, 3, 1, 4) Permutation 2: (4, 3, 2, 5, 1)

We can use these as interval numbers for X_1 (Permutation 1) and X_2 (Permutation 2). In order to get the specific values of X_1 and X_2, $n = 5$ random numbers are randomly selected from the standard uniform distribution. If we denote these values by U_m, where $m = 1, 2, 3, 4, 5$. Each random number U_m is scaled to obtain a cumulative probability P_m, so that each P_m lies within m-th interval:

$$P_m = \frac{U_m}{5} + \frac{m-1}{5} \tag{2.6}$$

In Tables 2.3 and 2.4, possible selections of Latin hypercube sample of size 5 for random variables X_1 and X_2 are presented respectively. Therefore if we apply the

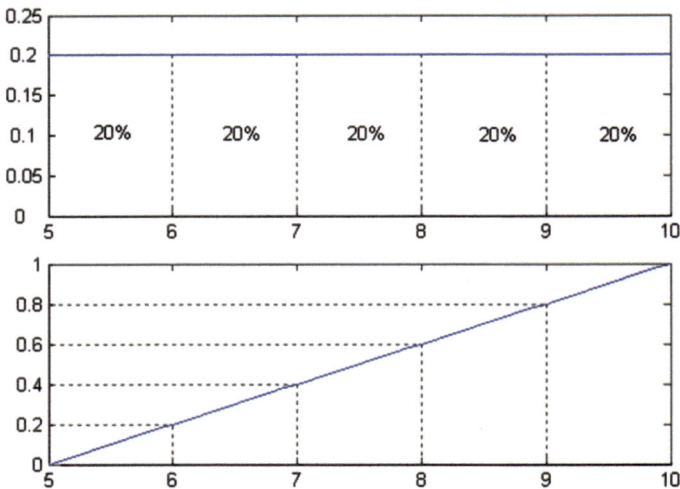

Fig. 2.9 Distribution and stratification for variable X_2

Table 2.3 Possible selection of values for a Latin hypercube sample of size 5 for the random variable X_1

Interval number (m)	Uniform (0,1) (U_m)	Scaled probabilities (P_m)	Corresponding sample
1	0.5832	0.1166	6.808
2	0.8125	0.3625	7.648
3	0.2980	0.4596	7.899
4	0.8470	0.7694	8.737
5	0.4369	0.8874	9.213

Table 2.4 Possible selection of values for a Latin hypercube sample of size 5 for the random variable X_2

Interval number (m)	Uniform (0,1) (U_m)	Scaled probabilities (P_m)	Corresponding sample
1	0.3370	0.0674	5.337
2	0.1678	0.2336	6.168
3	0.8419	0.5684	7.842
4	0.4372	0.6874	8.437
5	0.8127	0.9625	9.813

two permutations (Permutation 1 and 2) to choose the corresponding intervals for X_1 and X_2, as given in Table 2.5, we can perform the pairing operation. In Fig. 2.10, this pairing process is illustrated.

LHS was designed to improve the uniformity properties of Monte Carlo methods, since it was shown that the error of approximating a distribution by finite samples

Table 2.5 Pairing X_1 and X_2 and generating samples

Permutation 1 (interval for X_1)	Corresponding X_1	Permutation 2 (interval for X_2)	Corresponding X_2
2	7.648	4	8.437
5	9.213	3	7.842
3	7.899	2	6.168
1	6.808	5	9.813
4	8.737	1	5.337

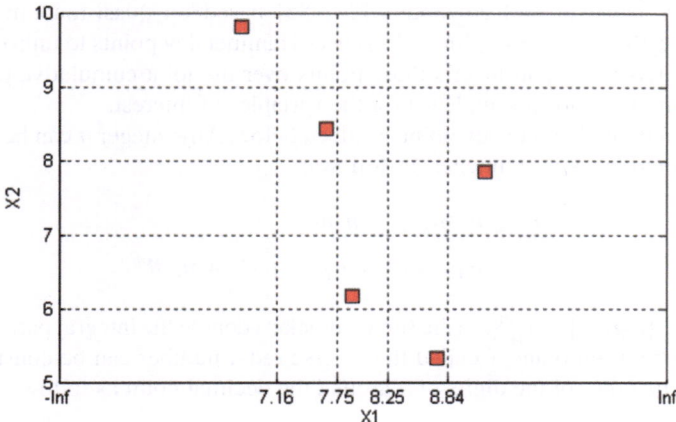

Fig. 2.10 Two-dimensional representation of a possible Latin hypercube sample of size 5 using X_1 and X_2

depends on the equidistribution properties of the sample used for U(0,1), and it is stated that the relationship between successive points in a sample or its randomness or independence is not critical [27]. In Median Latin Hypercube Sampling (MLHS), which is a variant of LHS, the mid-point of the intervals is chosen to sample the uncertain variables. MLHS is similar to the Descriptive Sampling described by [48].

   The main drawback of this stratification scheme in LHS and MLHS is that it is uniform in one dimension and does not provide uniformity properties in k-dimensions. Quasi-Monte Carlo methods can alleviate this problem and are described below.

2.3.3   Quasi-Monte Carlo Methods

Quasi-Monte Carlo methods seek to construct a sequence of points that perform significantly better than Monte Carlo, which has an average case of complexity of the order of $\frac{1}{\epsilon^2}$. For a suitably chosen set of samples, the quasi-Monte Carlo method provides a deterministic error bound of the order $n^{-1}(\log n)^{k-1}$ without any strong assumptions about the integrand. Some well-known quasi-Monte Carlo sequences are Halton, Hammersley, Sobol, Faure, Korobov and Neiderreiter [35]. The choice of an appropriate quasi-Monte Carlo sequence is a function of discrepancy. The deterministic upper and lower error bounds of any sequence for integration are expressed in terms of the discrepancy measure. Discrepancy is a quantitative measure for the deviation of the sequence from the uniform distribution. Therefore it is desirable to choose a low-discrepancy sequence. [15, 17] are some examples of low-discrepancy sequences.

Hammersley Sequence Sampling (HSS) is an efficient sampling technique developed by Diwekar and coworkers [24, 55] based on quasi-random numbers. Hammersley Sequence Sampling (HSS) uses Hammersley points to uniformly sample a unit hypercube and inverts these points over the joint cumulative probability distribution to provide a sample set for the variables of interest.

The design of Hammersley points is given below. Any integer n can be written in radix-R notation (R is an integer) as follows:

$$n \equiv n_m n_{m-1} \ldots n_1 n_0 \tag{2.7}$$

$$n = n_0 + n_1 R + n_2 R^2 + \cdots + n_m R^m \tag{2.8}$$

where $m = [\log_R n] = [\frac{ln(n)}{ln(R)}]$ (the square brackets denote the integral part). A unique fraction φ between 0 and 1 called the inverse radix number can be constructed by reversing the order of the digits of n around the decimal point as follows:

$$\varphi(n) = n_m n_{m-1} \ldots n_1 n_0 = n_0 R^{-1} + n_1 R^{-2} + \cdots + n_m R^{-(m+1)} \tag{2.9}$$

The Hammersley points on a k-dimensional cube are given by the following sequence:

$$\vec{Z_k}(n) = (n/N, \varphi_{R_1}(n), \varphi_{R_2}(n), \ldots, \varphi_{R_{k-1}}(n)) \tag{2.10}$$

where $R_1, R_2, \ldots, R_{k-1}$ are the first $k-1$ prime numbers. The Hammersley points are $\vec{x_k}(n) = 1 - \vec{Z_k}(n)$.

The following simple example illustrates how Hammersley points are generated.

Example 2.4 Generate 2-dimensional Hammersley points with a sample size of 100.

Solution In this case we have, $N = 100$ and $k = 2$. The $k-1$ prime numbers are $R_1 = 2$. The procedure for generating Hammersley points is given below for the first 10 points in Table 2.6.

Figure 2.11 shows the 100 points generated by HSS for $k = 2$.

As shown in the above example, the Hammersley sequence sampling (HSS) technique uses an optimal design scheme for placing n points on a k-dimensional hypercube. This scheme ensures that the samples are more representative of the population showing uniformity properties in multi dimensions, unlike Monte Carlo, Latin Hypercube, and its variant Median Latin Hypercube sampling techniques. A

Table 2.6 Generation of 10 Hammersley points in 2 dimensions

n	$\vec{Z_k}(n)$	2-Radix	$\varphi_2(n)$-inverse radix	$\vec{x_k}(n)$
0	$(0, \varphi_2(0))$	0	$\frac{0}{2^1} = 0$	$(1-0),(1-0)=(1.0,1.0)$
1	$(0.01, \varphi_2(1))$	1	$\frac{1}{2^1} = 0.5$	$(1-0.01),(1-0.5)=(0.99,0.5)$
2	$(0.02, \varphi_2(2))$	10	$\frac{0}{2^1} + \frac{1}{2^2} = 0.25$	$(1-0.02),(1-0.25)=(0.98,0.75)$
3	$(0.03, \varphi_2(3))$	11	$\frac{1}{2^1} + \frac{1}{2^2} = 0.75$	$(1-0.03),(1-0.75)=(0.97,0.25)$
4	$(0.04, \varphi_2(4))$	100	$\frac{0}{2^1} + \frac{0}{2^2} + \frac{1}{2^3} = 0.125$	$(1-0.04),(1-0.125)=(0.96,0.875)$
5	$(0.05, \varphi_2(5))$	101	$\frac{1}{2^1} + \frac{0}{2^2} + \frac{1}{2^3} = 0.625$	$(1-0.05),(1-0.625)=(0.95,0.375)$
6	$(0.06, \varphi_2(6))$	110	$\frac{0}{2^1} + \frac{1}{2^2} + \frac{1}{2^3} = 0.375$	$(1-0.06),(1-0.375)=(0.94,0.625)$
7	$(0.07, \varphi_2(7))$	111	$\frac{1}{2^1} + \frac{1}{2^2} + \frac{1}{2^3} = 0.875$	$(1-0.07),(1-0.875)=(0.93,0.125)$
8	$(0.08, \varphi_2(8))$	1000	$\frac{0}{2^1} + \frac{0}{2^2} + \frac{0}{2^3} + \frac{1}{2^4} = 0.0625$	$(1-0.08),(1-0.0625)=(0.92,0.9375)$
9	$(0.09, \varphi_2(9))$	1001	$\frac{1}{2^1} + \frac{0}{2^2} + \frac{0}{2^3} + \frac{1}{2^4} = 0.5625$	$(1-0.09),(1-0.5625)=(0.91,0.4375)$
10	$(0.10, \varphi_2(10))$	1010	$\frac{0}{2^1} + \frac{1}{2^2} + \frac{0}{2^3} + \frac{1}{2^4} = 0.3125$	$(1-0.10),(1-0.3125)=(0.90,0.6875)$

Fig. 2.11 Generation of 100 Hammersley points in 2 dimension

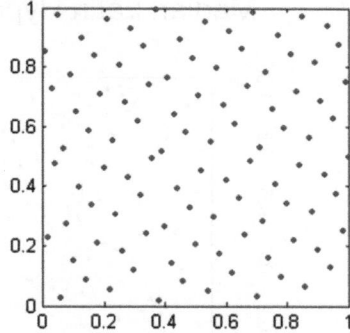

qualitative picture of the uniformity properties of the different sampling techniques on a unit square is presented in Fig. 2.12. It is clearly observed that HSS shows better uniformity than other stratified sampling techniques such as LHS, which are uniform along a single dimension only and do not guarantee a homogeneous distribution of points over the multivariate probability space.

One of the main advantages of Monte Carlo methods is that the number of samples required to obtain a given accuracy of estimates does not scale exponentially with the number of uncertain variables. HSS preserves this property of Monte Carlo. For correlated samples, the approach used by [24] uses rank correlations [22] to preserve stratified design along each dimension. Although this approach preserves the uniformity properties of the stratified schemes, the optimal location of the Hammersley points is perturbed by imposing the correlation structure. Figure 2.13 illustrates the effect of imposing a correlation structure on the sample sets. [24] have shown that

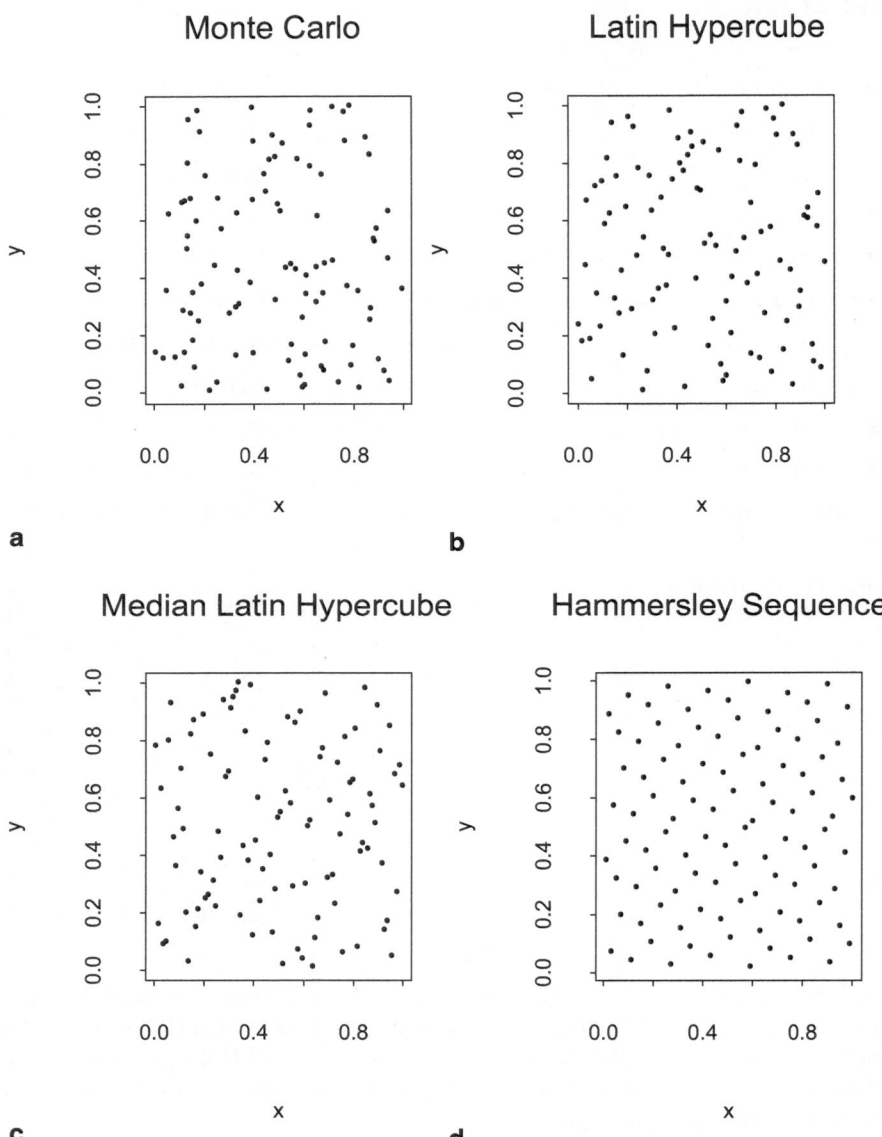

Fig. 2.12 Generation of 100 points on a unit square from various sampling techniques

HSS technique has better performance than LHS and crude Monte Carlo sampling techniques and is at least 3 to 100 times faster for convergence.

A variant of the HSS sampling technique is the Latin Hypercube Hammersley Sampling (LHSS) [60]. The aim of this sampling technique is to better utilize the 1-dimensional uniformity property of LHS and multidimensional uniformity property

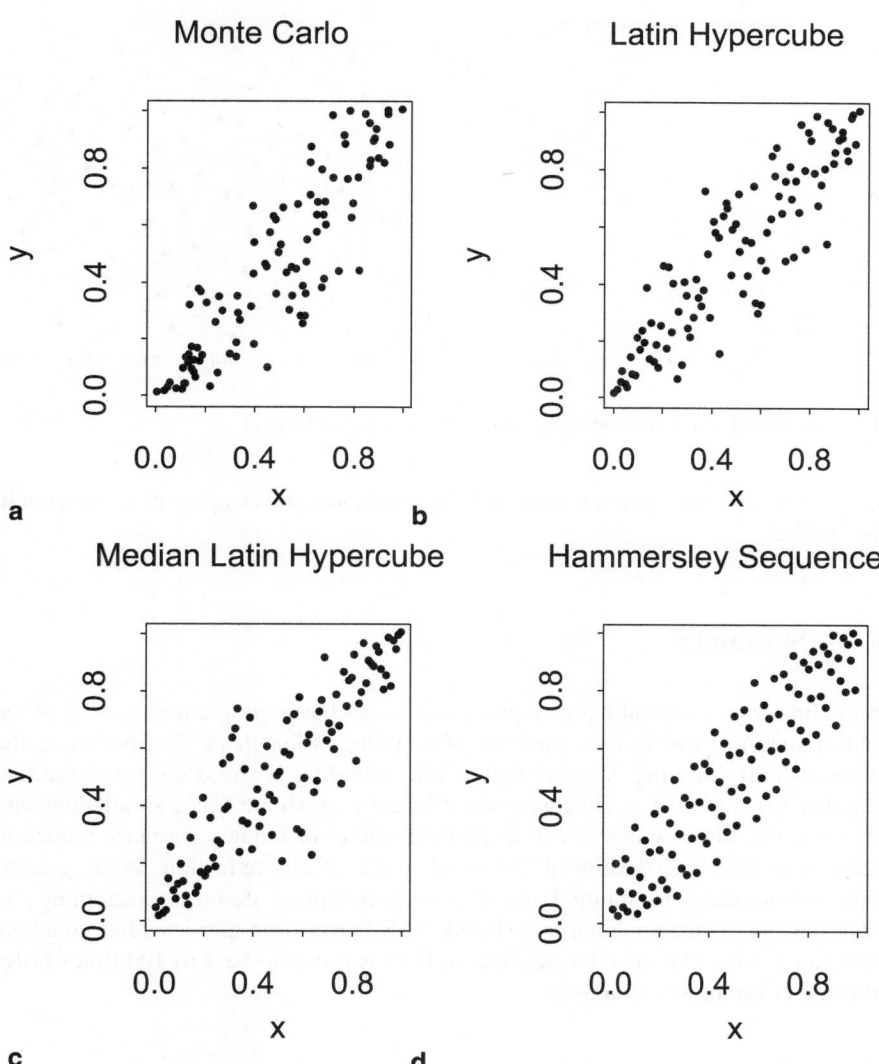

Fig. 2.13 Generation of 100 correlated points on a unit square from various sampling techniques

of HSS by coupling them. One dimensional uniformity analysis for Monte Carlo sampling, HSS, and LHSS is shown in Fig. 2.14. Other variants of Hammersley Sequence Sampling (HSS) are Halton sequence sampling or shifted Hammersley where the first variable is shifted, and leaped Halton or Hammersley, where some of the cycles of these sequences are eliminated to improve efficiency for higher dimensional problems [26, 55]. As the number of dimensions increase, the quasi-random sequences lose their uniformity properties. Therefore, to increase their performance,

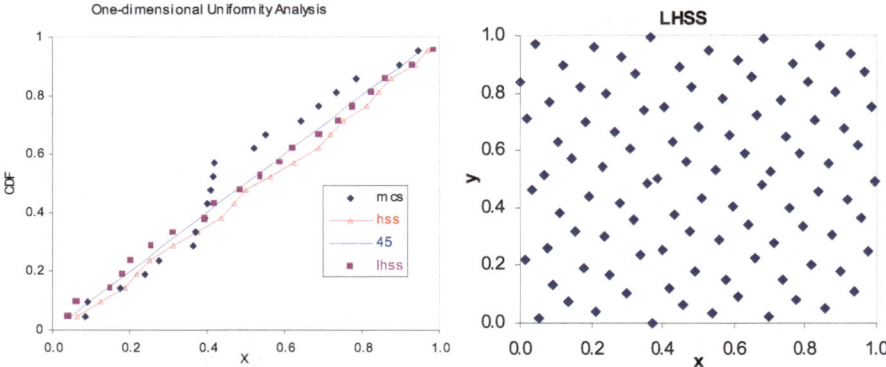

Fig. 2.14 One dimensional uniformity of various sampling techniques

different quasi-random sequences could be combined and a leaping procedure could be applied.

2.4 Summary

Sampling is an essential iterative procedure in stochastic programming. One of the oldest and most widely used methods of sampling probabilistic distributions is the Monte Carlo sampling. Crude Monte Carlo sampling is based on pseudorandom number generation. For increasing the efficiency of Monte Carlo simulations and to overcome disadvantages such as probabilistic error bounds, variance reduction techniques have been developed. Frequently used variance reduction sampling methods are importance sampling, Latin Hypercube Sampling, descriptive sampling and Hammersley Sequence Sampling (HSS). HSS is based on quasi-random numbers generated using Hammersley sequences. HSS is found to be 3 to 100 times faster than other sampling techniques.

Notations

a	multiplier in Lehmer linear congruent generator
c	increment in Lehmer linear congruent generator
$F()$	function
$G()$	biased distribution
I	integral
I_n	n^{th} random number from Lehmer linear congruent generator
m	modulus in Lehmer linear congruential generator $[\log_R n] = [\frac{ln(n)}{ln(R)}]$ for HSS

P_m scaled probabilities
R integer in R-radix notation
x_i, X_i random number
$\overrightarrow{x_k}(n)$ Hammersley points
U_m samples from uniform distribution (U(0,1))

Greek letters
ϵ error
σ standard deviation
$\varphi(n)$ inverse radix function for n

Chapter 3
Probability Density Functions and Kernel Density Estimation

Stochastic modeling loop in the stochastic optimization framework involves dealing with evaluation of a probabilistic objective function and constraints from the output data. Probability density functions (PDFs) are a fundamental tool used to characterize uncertain data. Equation 3.1 shows the definition of a PDF f of variable X.

$$P(a \leq X \leq b) = \int_a^b f(x)dx \qquad (3.1)$$

We all are familiar with the PDF for distributions like normal distribution where parameters like the mean and the variance can be used to define the distribution. However, when dealing with the generalized case of PDF, we may not be able to categorize it in terms of parametric distributions like normal or lognormal. For these cases, we have to depend on nonparametric approach for estimating PDF.

3.1 The Histogram

The oldest and widest used method of nonparametric density estimation is the histogram. A histogram is constructed by dividing the data into intervals of bins and counting the frequency of points in that bin. Given an origin and a bin width h, the histogram can be defined by the following function (Eq. 3.2).

$$f(x) = \frac{1}{nh} \ (no.\ of\ X_i\ in\ the\ same\ bin\ as\ x), \qquad (3.2)$$

where n is the total observations.

In a histogram, the important parameter to be chosen is the bin width h. Figure 3.1 shows a typical histogram. An estimator for a histogram can be written as

$$f(x) = \frac{1}{n} \sum_i^n \frac{1}{h} w \left(\frac{x - X_i}{h} \right) \qquad (3.3)$$

© Urmila Diwekar, Amy David 2015

U. Diwekar, A. David, *BONUS Algorithm for Large Scale Stochastic Nonlinear Programming Problems*, SpringerBriefs in Optimization, DOI 10.1007/978-1-4939-2282-6_3

Fig. 3.1 A typical histogram

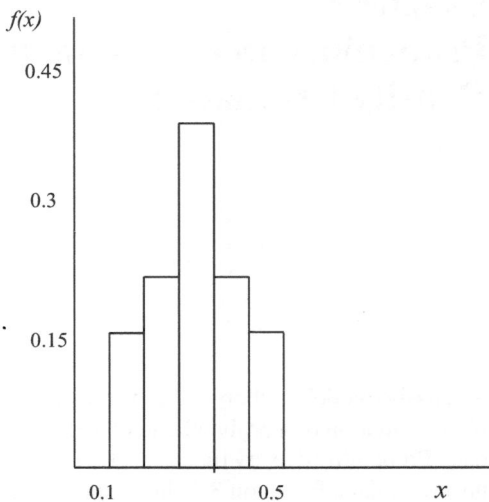

where w is the weight function defined in terms of any variable y by

$$w(y) = \frac{1}{2} \quad if \quad |y| < 1$$
$$= 0 \quad otherwise \tag{3.4}$$

Although histograms are very useful, it is difficult to represent bivariate or trivariate data with histograms. Further, it can be seen that histogram does not represent a continuous function and requires smoothing. Kernel density estimation (KDE) overcomes these difficulties with histograms and is the focus of this chapter. This chapter is based on the book by Silverman [52].

3.2   Kernel Density Estimator

The kernel density estimator with Kernel K is defined by

$$\int_{-\inf}^{\inf} K(x)dx = 1 \tag{3.5}$$

or

$$f(x) = \frac{1}{nh} \sum_{i=1}^{n} K\left(\frac{x - X_i}{h}\right), \tag{3.6}$$

where h is the window width, also called the smoothing parameter or bandwidth.

The multivariate KDE for d dimensions is given by

$$f(x) = \frac{1}{nh^d} \sum_{i=1}^{n} K\left(\frac{x - X_i}{h}\right), \tag{3.7}$$

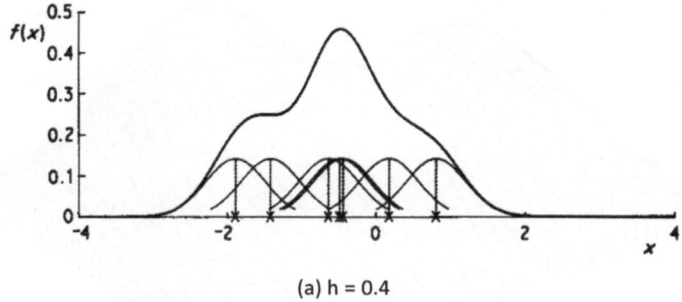

(a) h = 0.4

Fig. 3.2 Probability density function from a normal KDE. *KDE* kernel density estimation

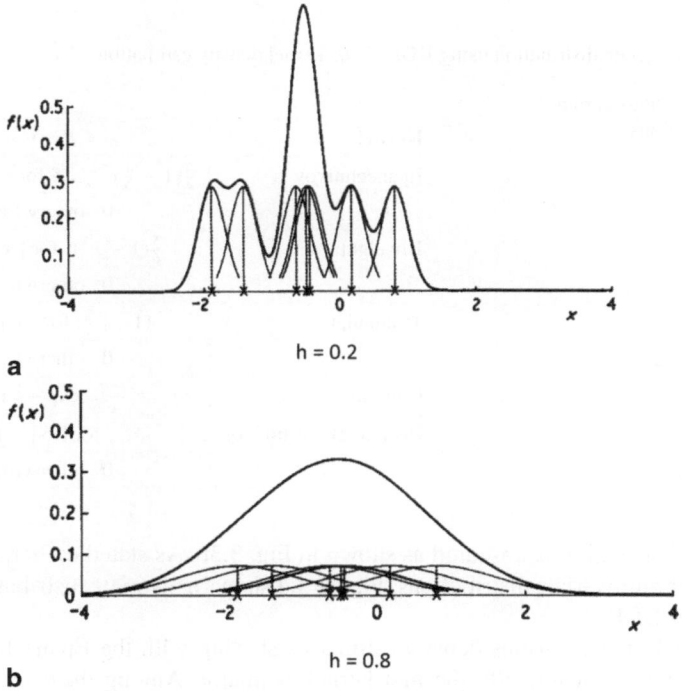

h = 0.2

a

h = 0.8

b

Fig. 3.3 Effect of *h* on PDF. *PDF* probability density function

If K is a generally radially symmetric unimodal function like the normal density function then the PDF f will be a smooth curve and derivatives of all orders can be calculated. This is important for optimization algorithms. Figure 3.2 shows the density function derived from normal density functions. Just like histograms that are considered sums of boxes, Kernel estimator is considered sums of bumps. The smoothing parameter h is very important in KDE. If h is too small then spurious structures result as shown in Fig. 3.3a. For too large a choice of h, the bimodal

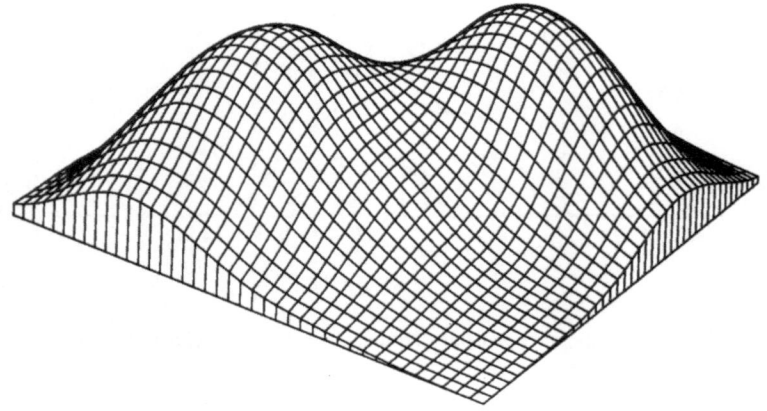

Fig. 3.4 A bivariate distribution using KDE. *KDE* kernel density estimation

Table 3.1 Various kernel density functions

Kernel	$K(x)$
Epanechnikov	$\frac{3}{4}(1 - \frac{1}{5}x^2)/\sqrt{5}$ for $\mid x \mid < \sqrt{5}$,
	0 otherwise
Biweight	$\frac{5}{6}(1 - x^2)^2$ for $\mid x \mid < 1$,
	0 otherwise
Triangular	$(1- \mid x)$ for $\mid x \mid < 1$,
	0 otherwise
Gaussian	$\frac{1}{\sqrt{2\pi}} \exp -\frac{1}{2}x^2$
Rectangular estimator	$\frac{1}{2}$ for $\mid x \mid < 1$,
	0 otherwise

nature of distribution is obscured as shown in Fig. 3.3b. As stated earlier, KDE can be used for multivariate distributions. Figure 3.4 shows a bivariate distribution using a Gaussian KDE.

Table 3.1 shows various density estimators starting with the Epanechnikov estimator which is historically the first kernel estimator. Among these estimators a Gaussian or a normal estimator is commonly used. Therefore, we will be focusing on normal kernel estimator for deriving the BONUS algorithm.

The optimal smoothing parameter h_{opt} for the Gaussian KDE is given by

$$h_{opt} = 1.06\sigma n^{-\frac{1}{5}}, \tag{3.8}$$

where σ is the standard deviation of the observations and n is the number of observations.

The following example shows how a probability density function based on a Gaussian KDE is estimated.

Table 3.2 Values of decision variables and objective function for ten samples

Sample No.	x_1	x_2	Z
1	5.6091	0.3573	15.2035
2	3.7217	1.9974	14.7576
3	6.2927	4.2713	0.5738
4	7.2671	3.3062	0.5527
5	4.1182	1.3274	15.4478
6	7.7831	1.5233	6.7472
7	6.9578	1.1575	8.0818
8	5.4475	3.6813	2.5119
9	8.8302	2.9210	4.5137
10	6.9428	3.7507	0.0654
Mean	6.2970	2.4293	–
Standard deviation	1.5984	1.3271	–

Example 3.1 Consider the following optimization problem. We use the normal distribution for the uncertain parameters, a uniform distribution for decision variables, and uniform distributions using the minimum and maximum values of the decision variable to generate the data for the objective function Z. These values are given in Table 3.2. Find the PDF, f, using a Gausian KDE.

$$\min E[Z] = E[(\tilde{x}_1 - 7)^2 + (\tilde{x}_2 - 4)^2] \tag{3.9}$$

$$\text{s.t. } \tilde{x}_1 \in N[\mu = x_1^\star, \sigma = 0.033 \cdot x_1^\star] \tag{3.10}$$

$$\tilde{x}_2 \in U[0.9 \times x_2^\star, 1.2 \times x_2^\star] \tag{3.11}$$

$$4 \le x_1 \le 10 \tag{3.12}$$

$$0 \le x_2 \le 5 \tag{3.13}$$

Here, E represents the expected value, and the goal is to minimize the mean of the objective function Z calculated for two uncertain decision variables, x_1 and x_2.

Solution The Gaussian KDE formula for $f(x)$ is given below, where X_i is the value of ith observation.

$$f(x) = \frac{1}{\sqrt{2\pi}nh} \sum_{i=1}^{n} \exp -\frac{1}{2}\left(\frac{x - X_i}{h}\right)^2 \tag{3.14}$$

From the values of observations for variable Z given in Table 3.2, the value of standard deviation σ is found to be 6.29917. Therefore, using Eq. 3.8 we can find the value of optimum h to be

$$h = 1.06 \times 6.29917 \times 10^{-\frac{1}{5}}$$

$$= 4.212978 \tag{3.15}$$

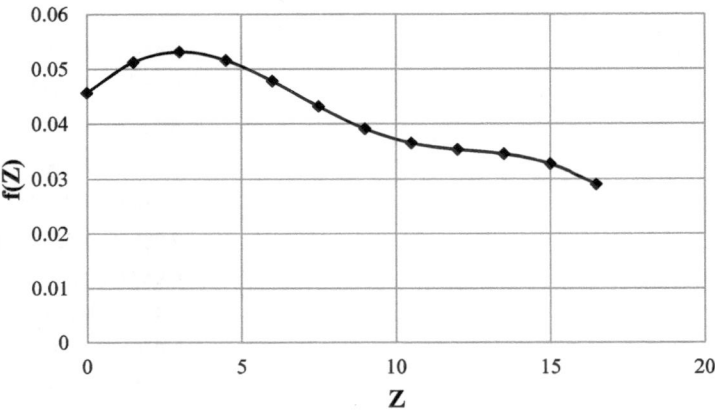

Fig. 3.5 PDF for the output variable Z. *PDF* probability density function

In order to obtain $f(Z)$, we have divided the region from $Z = 0$ to $Z = 15$. The value of PDF $f(Z)$ is calculated using Eq. 3.14 for various values of Z as shown in Table 3.2.

Figure 3.5 shows the PDF for Z obtained using Gaussian KDE (Table 3.3).

3.3   Summary

Kernel density estimation provides a nonparametric way to estimate probability density function. Symmetric and unimodal KDE functions like normal KDE provides a continuous smooth function where derivatives can be estimated. A Gaussian KDE is commonly used for this purpose. The value of smoothing parameter h is important in KDE. If h is too small then spurious structures result and if h is too large then the real nature of the probability density function is obscured. The optimal value of smoothing parameter is a function of number of observations and standard deviation of distribution. A Gaussian KDE provides a basis for the BONUS algorithm.

Table 3.3 Kernel density estimation for the objective function Z

Z	15.2035	14.7576	0.5738	0.5527	$\exp-\frac{1}{2}(\frac{Z-Z_i}{h})^2$ 15.4478	6.7472	8.0818	2.5119	4.5137	0.0654	$f(Z)$
0	0.00149	0.00217	0.99077	0.99143	0.00120	0.27736	0.15882	0.83716	0.56331	0.99988	0.045688
1.5	0.00504	0.00707	0.97612	0.97504	0.00417	0.46042	0.29513	0.97157	0.77426	0.94367	0.051266
3	0.01507	0.02036	0.84720	0.84475	0.01272	0.67331	0.48312	0.99331	0.93749	0.78459	0.053155
4.5	0.03966	0.05161	0.64775	0.64472	0.03417	0.86740	0.69670	0.89463	0.99999	0.57465	0.051633
6	0.09198	0.11526	0.43629	0.43348	0.08090	0.98440	0.88507	0.70982	0.93967	0.37078	0.04781
7.5	0.18792	0.22677	0.25888	0.25675	0.16873	0.98416	0.99051	0.49613	0.77785	0.21075	0.043177
9	0.33821	0.39304	0.13532	0.13397	0.31001	0.86678	0.97653	0.30549	0.56724	0.10553	0.039138
10.5	0.53622	0.60011	0.06231	0.06158	0.50176	0.67251	0.84812	0.16571	0.36440	0.04655	0.036554
12	0.74894	0.80717	0.02528	0.02494	0.71543	0.45966	0.64890	0.07918	0.20622	0.01809	0.035366
13.5	0.92150	0.95642	0.00903	0.00890	0.89864	0.27677	0.43736	0.03333	0.10281	0.00619	0.034581
15	0.99883	0.99835	0.00284	0.00280	0.99437	0.14681	0.25969	0.01236	0.04515	0.00187	0.032801
16.5	0.95375	0.91803	0.00079	0.00077	0.96929	0.06860	0.13583	0.00404	0.01747	0.00050	0.02907

Notations

d	dimension
E	expected value function
f	probability density function
h	bin width
h_{opt}	optimum bin width
K	kernel density function
n	number of observations
w	weight function
Z	output variable

Greek letters

σ	standard deviation

Chapter 4
The BONUS Algorithm

In this chapter we describe the basics of the Better Optimization of Nonlinear Uncertain System (BONUS) algorithm. For better readability, we present the generalized stochastic optimization framework (Fig. 1.4 (from Chap. 1) for stochastic nonlinear programming (NLP) problem below. This chapter is derived from the work by [43].

General techniques for these types of optimization problems determine a statistical representation of the objective such as maximum expected value or minimum variance. Once embedded in an optimization framework, the iterative loop structure emerges where decision variables are determined, a sample set based on these decision variables is generated, the model is evaluated for each of these sample points, and the probabilistic objective function value and constraints are evaluated, as shown in the inner loop of the Fig. 4.1. When one considers that nonlinear optimization techniques rely on an objective function and constraints evaluation for each iteration, along with derivative estimation through perturbation analysis, the sheer number of model evaluations rises significantly rendering this approach ineffective for even moderately complex models. Figure 4.2 shows the general idea behind the BONUS algorithm. BONUS follows the grey arrows. In the stochastic optimization iterations (Fig. 4.1), decision variables values can vary between upper and lower bounds, and in sampling loop various probability distributions are assigned to uncertain variables. In the BONUS approach, initial uniform distributions (between upper and lower bounds) are assumed for decision variables. These uniform distributions together with specified probability distributions of uncertain variables form the base distributions for analysis. BONUS samples the solution space of the objective function at the beginning of the analysis by using the base distributions. As decision variables change, the underlying distributions for the objective function and constraints change, and the proposed algorithm estimates the objective function and constraints values based on the ratios of the probabilities for the current and the base distributions (a reweighting scheme), which are approximated using kernel density estimation (KDE) techniques. Thus, BONUS avoids sample model runs in subsequent iterations.

© Urmila Diwekar, Amy David 2015

U. Diwekar, A. David, *BONUS Algorithm for Large Scale Stochastic Nonlinear Programming Problems*, SpringerBriefs in Optimization, DOI 10.1007/978-1-4939-2282-6_4

Fig. 4.1 Pictorial
representation of the
stochastic programming
framework

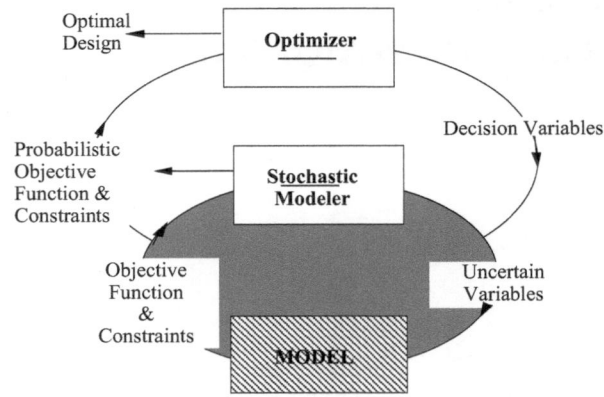

4.1   Reweighting Schemes

The goal of the reweighting scheme (shown by gray arrows in Fig. 4.2) is to determine changes in output distributions as input distributions change. Hesterberg
(1995) presents various reweighting techniques for estimating the expected value of
an output distribution cumulative distribution function (CDF), $F[J(u)]$ without evaluating the model for the input distribution probability density function (PDF), $f(u)$)
in Fig. 4.2. The ratio of the probability density functions f is used as a weight, which
is given as:

$$\omega_i = \frac{f(u_i)}{\hat{f}(u_i^\star)}, \tag{4.1}$$

where $\hat{f}(u_i^\star)$ is determined for the base sample set, for which the model response is
known, and the probability density $f(u_i)$ is calculated using the sample for which
the response has to be estimated. Remember that these two sample sets are not
necessarily related. One attempt for estimating statistical properties $P(u)$ for the
output of the model is through the product of the weights and the same properties
obtained from the base distribution (Eq. 4.2).

$$P(u) = \sum_i \omega_i \cdot P(u_i^\star) \tag{4.2}$$

For instance, to estimate the mean μ of a model response, $Z(u)$, the weight would
be multiplied by the individual model responses for the base set:

$$\mu[Z(u)] = \sum_i^{N_{samp}} \omega_i \cdot Z(u_i^\star), \tag{4.3}$$

where N_{samp} is the sample size.

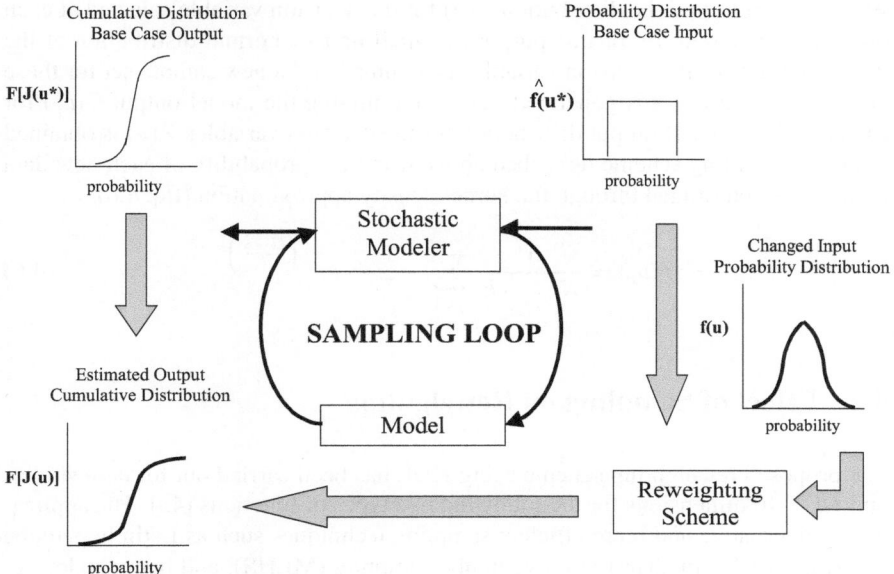

Fig. 4.2 Density estimation approach to optimization under uncertainty

This approach has limitations, as the weights may not sum to 1. This problem is reduced by using normalized weights, as shown in Eq. 4.4. This normalized reweighting (the ratio estimate for weighted average) has another advantage as it provides acceptable performance for a wider range of perturbation, especially for large samples of Monte Carlo simulations (MCS) [19]. In the BONUS, instead of using large size of MCS, a more efficient sampling technique as presented in Chap. 2 that provides the same accuracy as MCS in order of magnitude with less number of samples is used.

$$P(u) = \sum_{j}^{N_{samp}} \frac{\frac{f(u_j)}{\hat{f}(u_j^\star)}}{\sum_{i=1}^{N_{samp}} \frac{f(u_i)}{\hat{f}(u_i^\star)}} \cdot P(u_j^\star) \tag{4.4}$$

As seen in Eq. 4.4 the mean of the function can be estimated from the ratio of the two input distributions $f(u)$ and $\hat{f}(u^\star)$. This requires the determination of the probability distributions from a given sample set of uncertain variables. Here, the KDE techniques discussed in Chap. 3 are used.

In order to use the kernel density approach for estimating function values (objective function and constraints), the base sample set $u^\star$ has to be generated for model calculations. As stated earlier, we select uniform distributions for the decision variables and specified distributions for uncertain variables for creating the base sample. Once the base sample is obtained, its density can be calculated for each point as:

$$\hat{f}(u_i^\star) = \frac{1}{N_{samp} \cdot h} \sum_{j=1}^{N_{samp}} \frac{1}{\sqrt{2\pi}} \cdot e^{-\frac{1}{2}\left(\frac{u_i^\star - u_j^\star}{h}\right)^2} \tag{4.5}$$

We now want to find the distribution $f(u)$ for the decision variable selected at each optimization iteration. For this purpose a small narrow normal distribution at the decision point for the decision variables is assumed and a new sample set for these normal distributions u is generated. After determining the model output $Z(u_i^\star)$ for each $u_i^\star$, the value of output distribution for the decision variables $Z(u)$ is obtained by the reweighting scheme described above using the probability of each new data point u_i, as determined through the kernel density approximation (Eq. 4.6).

$$f(u_i) = \frac{1}{N_{samp} \cdot h} \sum_{j=1}^{N_{samp}} \frac{1}{\sqrt{2\pi}} \cdot e^{-\frac{1}{2}\left(\frac{u_i - u_j^\star}{h}\right)^2} \tag{4.6}$$

4.2   Effect of Sampling on Reweighting

The proposed reweighting scheme using KDE has been carried out for case studies up to $d = 10$ dimensions for the following five types of functions [43]. The application of alternative and more efficient sampling techniques such as Latin hypercube sampling (LHS), median Latin hypercube sampling (MLHS), and hammersley sequence sampling (HSS) have resulted in significant reductions of computational requirements compared to MCS as shown in this section.

- Function 1: Linear additive: $y = \sum_{m=1}^{s} u_m$ $s = 2...10$
- Function 2: Multiplicative: $y = \Pi_{m=1}^{s} u_m$ $s = 2...10$
- Function 3: Quadratic: $y = \sum_{m=1}^{s} u_m^2$ $s = 2 \dots 10$
- Function 4: Exponential: $y = \sum_{m=1}^{s} u_m \cdot exp(u_m)$ $s = 2 \dots 10$
- Function 5: Logarithmic: $y = \sum_{m=1}^{s} log(u_m)$ $s = 2 \dots 10$

The total analysis includes five functions, with four sampling techniques being compared for each of these functions. The number of sample points for each sample is also analyzed, by selecting sample sizes as $N_{samp} = [50, 100, 250, 500, 750, 1000, 2500, 5000, 7500, 10,000]$. This results in a total of 200 runs for which the proposed reweighting approach has been tested. For each run, the means and variances are both calculated and estimated, as are the derivatives of each of these with respect to each u. Further, the percentage error between the actual and estimated values is determined as well, as shown in Table 4.1.

As required, the base distributions are uniform distributions of decision variables with bounds given in first three columns of Table 4.2, and the estimated distributions were narrow normal, with the upper and lower bounds in the last three columns of the table indicating the region enclosing the 99.999 percentile.

For the generation of the shifted sample set u^Δ and for derivative calculations, the step size Δu_j was selected as:

$$\Delta u_j = 0.05 \cdot \mu\{u_j\} \tag{4.7}$$

As the model functions are relatively simple, the actual values (analytical) of the mean and variance for both sample sets u and u^Δ are calculated, and compared to

Table 4.1 Calculations for KDE efficiency analysis

n-dimensional calculations	2 to 10 dimensions = 10! = 3628800
Functions	5
Sampling techniques	4
Sample sizes	10
Total runs	$5 \times 4 \times 10 = 200$
Moment calculations/run	4
Derivative calculations/run	$10! \times 2$
Moment estimations/run	4
Derivative estimations/run	$10! \times 2$
% Error calculations moments/run	2
% Error calculations derivatives/run	$10! \times 2$
Total calculations	$(4 + 4 + 2 + 6 \times 10!) \times 200 \simeq 4.32 \times 10^9$

Table 4.2 Bounds for base (uniform) and estimated (normal) distributions

	Base distribution			Estimated distribution	
	Lower bound	Upper bound		Lower bound	Upper bound
$u_1^\star$	1.0	6.0	u_1	3.0	5.0
$u_2^\star$	3.0	7.0	u_2	4.0	7.0
$u_3^\star$	1.0	5.0	u_3	3.0	4.0
$u_4^\star$	8.0	12.0	u_4	9.5	10.0
$u_5^\star$	10.0	17.0	u_5	11.5	14.0
$u_6^\star$	2.0	9.0	u_6	4.0	6.0
$u_7^\star$	3.0	7.0	u_7	4.5	6.5
$u_8^\star$	0.0	7.5	u_8	1.0	6.0
$u_9^\star$	10^{-5}	10^{-1}	u_9	5×10^{-3}	5×10^{-2}
$u_{10}^\star$	6.0	9.0	u_{10}	8.0	9.0

the estimates. Further, the same analysis is conducted for the derivative estimates, allowing for comparison of the errors in the estimates based on the sampling technique that is applied to generate both sample sets $u^\star$ and u. The next section provides the results of the preliminary study.

As indicated above, 200 different runs have been used to verify the applicability of the technique. For each run, means, variances, and derivatives have been calculated and estimated using the reweighting scheme, and percentage errors between each of these have been determined. Due to the extensive nature of this analysis, only one example is provided here that is both relevant to this analysis as well as representative of the overall behavior of the technique.

The results obtained for the nonlinear function, $y = \sum_{m=1}^{3} u_m^2$ are presented here. Variance calculation is more prone to errors than calculation of mean (if sample size

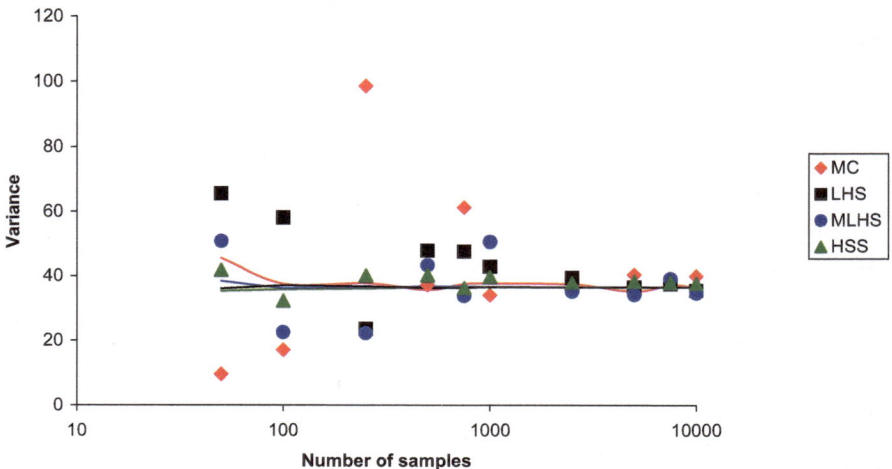

Fig. 4.3 Variance calculation for different sampling techniques

Table 4.3 Percentage error in variance estimation for 3-dimensional analysis using 250 samples

	MCS	LHS	MLHS	HSS
Function 1	178.5688	34.6615	50.0478	7.3945
Function 2	179.7385	30.1337	54.5286	11.1636
Function 3	161.1127	36.0205	39.1106	10.9293
Function 4	140.1933	9.2476	13.0681	4.1226
Function 5	183.3928	30.1835	54.2601	8.9386

is small), and also the case study in the next section aims at calculating the variance of the system at hand that presents the efficiency of the reweighting technique to estimate variance for this function here.

Simultaneous plotting of the actual and estimated values will allow one to identify how accurate each technique is. Note that the x-axis is in log scale to capture the change of the sample sizes through N_{samp} = [50, 100, 250, 500, 750, 1000, 2500, 5000, 7500, 10,000]. The lines represent the actual values, while the stand-alone points represent the estimated variance values using the four different sampling techniques.

In Fig. 4.3, the variance of Function 3 is plotted with respect to the number of samples. As seen, all four sampling techniques converge to the same value as N_{samp} approaches 10,000, with the MCS technique showing the highest variations. While most approaches over- or underestimate the mean at low sample sizes, HSS provides a rather accurate estimate in this region. Table 4.3 provides the percentage error between the estimates and the actual values of the variance for $f(u)$ for all four

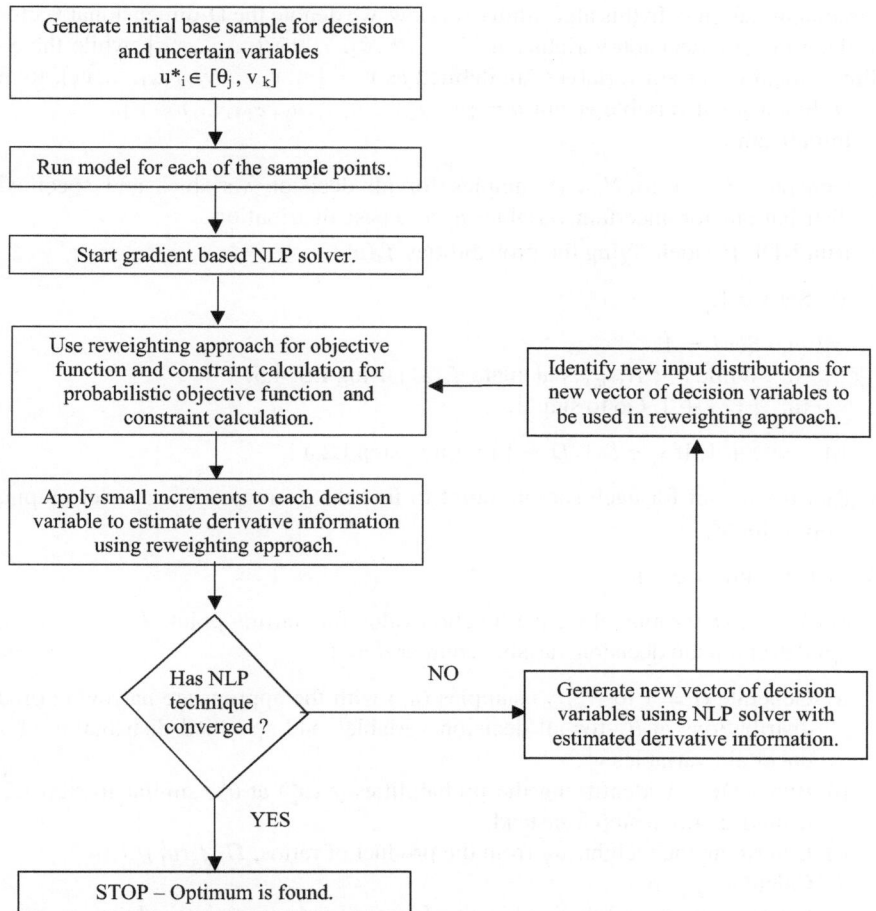

Fig. 4.4 Optimization under uncertainty: The BONUS algorithm

sampling techniques with sample sizes of 250. As seen, HSS yields comparably small percentage errors for all functions.

4.3   BONUS: The Novel SNLP Algorithm

The algorithm for BONUS, given in Fig. 4.4, can be divided into two sections. The first section, Initialization, starts with generating the base distribution that will be used as the source for all estimations throughout the optimization. After the base distribution is generated, the second section starts, which includes the estimation technique that results in the improvements associated with BONUS with respect to computational time. In this algorithm overview, we denote the D-dimensional vector of deterministic decision variables as $\theta = [\theta_1, \theta_2, ..., \theta_d, \theta_{d+1}, .., x_D]$, while the S-dimensional uncertain variables are defined as $v = [v_1, v_2, ..., v_s, v_{s+1}, ..., v_S]$, total S+D-dimensional variable vector $u = [u_1, u_2, ..., u_{s+d}, u_{s+d+1}, ..., u_{S+D}]$..

I - Initialization

1. Generate ($i = 1$ to N_{samp}) samples for all decision variables and specified distributions for uncertain variables $u_i^\star$ as a base distribution.

2. Run KDE for identifying the probabilities $\hat{f}_s(u_i^\star)$.

   a) Set $s = 1$.

      i. Set $i = 1$.
      ii. While $i < N_{samp}$, calculate $\hat{f}_s(u_i^\star)$ using Eq. 4.5.
      iii. $i = i + 1$. Go to step ii.

   b) $s = s + 1$. If $s < S + D + 1$ return to step I.2.a.i.

3. Run the model for each sample point to find the corresponding model output, store value Z_i.

II - SNLP Optimization

1. Set $k = 1$. Determine objective function value for starting point, $J = P(\theta^k, v^k)$. Set deterministic decision variable counter $d = 1$.

   a) Generate ($i = 1$ to N_{samp}) samples (u_i^k) with the appropriate narrow normal distributions at θ_d^k for all decision variables and specified distributions for uncertain variables v_i^k.

   b) Run KDE for identifying the probabilities $f_s(u_i^k)$ at θ_d^k, similar to step I.2, using Eq. 4.6 in step ii instead.

   c) Determine the weights ω_i from the product of ratios, $\Pi_S f_s(u_i^k)/\hat{f}_s(u_i^\star)$.

   d) Calculate $\sum_i \omega_i$.

   e) Estimate the probabilistic objective function and constraints values:

      i. Set $i = 1$, $J^k = 0$.
      ii. While $i < N_{samp}$, calculate: $J^k = J_i^k * \omega_i / \sum_i \omega_i$.
      iii. $i = i + 1$. Go to step ii.

   f) Set $d = d + 1$, return to step II.2.

2. While $d \leq D$, perturb one decision variable θ_d^k to find $\theta_d^{k,\Delta}$. Reset deterministic decision variable counter $d = 1$.

   a) Generate ($i = 1$ to N_{samp}) samples with the appropriate distributions at $\theta_d^{k,\Delta}$ for all variables u_i^k.

   b) Run KDE for identifying the probabilities $f_s(u_i^k)$ at $\theta_d^{k,\Delta}$, similar to steps I.2, using Eq. 4.6 in step ii instead.

   c) Determine the weights ω_i from the product of ratios, $\Pi_S f_s(u_i^k)/\hat{f}_s(u_i^\star)$.

   d) Calculate $\sum_i \omega_i$.

e) Estimate probabilistic objective function and constraints value:

    i.   Set $i = 1$, $J^{k,\Delta} = 0$.
    ii. While $i < N_{samp}$, calculate: $J^{k,\Delta} = J_i^{k,\Delta} * \omega_i / \sum_i \omega_i$.
    iii. $i = i + 1$. Go to step ii.

f) Set $d = d + 1$, return to step II.2.

3. Calculate gradient information obtained from II-1 and II-3.
4. Check convergence criteria for nonlinear solver (KKT conditions); if satis-
    fied, STOP-Optimum found. Otherwise, identify new vector of decision vari-
    ables through gradients obtained from objective function value estimation via
    reweighting. Set $k = k + 1$. Return to step II-2.

Note that traditional techniques rely on repeated model runs for steps II-3b in the
algorithm. For computationally complex nonlinear models, this task can become the
critical bottleneck for solving the SNLP. BONUS, on the other hand, bypasses these
by estimating the objective function values via reweighting. The BONUS algorithm is
implemented using the nonlinear solver based on sequential quadratic programming
(SQP) method. The following examples illustrate the steps involved in BONUS and
the efficiency of BONUS for solving SNLP problems.

Example 4.1   Consider the optimization problem presented in Example 3.1 again.
Illustrate the reweighting scheme and solve the problem using BONUS.

$$\min E[Z] = E[(\tilde{x}_1 - 7)^2 + (\tilde{x}_2 - 4)^2] \tag{4.8}$$

$$\text{s.t. } \tilde{x}_1 \in N[\mu = x_1^\star, \sigma = 0.033 \cdot x_1^\star] \tag{4.9}$$

$$\tilde{x}_2 \in U[0.9 \cdot x_2^\star, 1.2 \cdot x_2^\star] \tag{4.10}$$

$$4 \le x_1 \le 10 \tag{4.11}$$

$$0 \le x_2 \le 5 \tag{4.12}$$

Here, E represents the expected value, and the goal is to minimize the mean of the
objective function calculated for two uncertain decision variables, x_1 and x_2. The
optimizer determines the value $x_1^\star$, which has an underlying normal distribution with
$\pm 10\%$ of the nominal value of $x_1^\star$ as the upper and lower 0.1 % quantiles. Similarly,
$\tilde{x}_2$ is uniformly distributed around $x_2^\star$, with cutoff ranges at $[-10\%, +20\%]$.

Solution   The following steps illustrate the steps of BONUS algorithm to solve this
problem.

Step 1   The first step in BONUS is determining the base distributions for the decision
variables and uncertain variables, followed by generating the output values for this
model. Since in this case decision variable and uncertain variables are merged, we
use the entire possible range for the two variables as these base distributions have to
cover the entire range, including variations. For instance, for x_2, the range extends to
$(0 \times 0.9) \le x_2 \le (5 \times 1.2)$ to account for the uniformly distributed uncertainty. Due to
space limitations, the illustrative presentation of the kernel density and reweighting

Table 4.4 Base sample

Sample no.	x_1	x_2	Z
1	5.6091	0.3573	15.2035
2	3.7217	1.9974	14.7576
3	6.2927	4.2713	0.5738
4	7.2671	3.3062	0.5527
5	4.1182	1.3274	15.4478
6	7.7831	1.5233	6.7472
7	6.9578	1.1575	8.0818
8	5.4475	3.6813	2.5119
9	8.8302	2.9210	4.5137
10	6.9428	3.7507	0.0654
Mean	6.2970	2.4293	–
Std. Dev	1.5984	1.3271	–

approach is performed for a sample size of 10, while the remainder of the work uses $N = 100$ samples. A sample realization using MCS is given in Table 4.4.

After this sample is generated, KDE for the base sample is applied to determine the probability of each sample point with respect to the sample set. This is performed for each decision variable separately by approximating each point through a Gaussian kernel, and adding these kernels to generate the probability distribution for each point, as given in Eq. 4.13 [52].

$$
\hat{f}(x_i(k)) = \frac{1}{N \cdot h} \sum_{j=1}^{N} \frac{1}{\sqrt{2\pi}} \cdot e^{-\frac{1}{2}\left(\frac{x_i(k)-x_i(j)}{h}\right)^2}. \tag{4.13}
$$

Here, h is the width for the Gaussian kernel and depends on the variance σ and sample size N of the data set and is given as follows:

$$
h = 1.06 \times \sigma \times N^{-\frac{1}{5}}. \tag{4.14}
$$

For our example, $h(x_1) = 1.06 \times 1.5984 \times 10^{-0.2} = 1.0690$ and $h(x_2) = 1.06 \times 1.3271 \times 10^{-0.2} = 0.8876$. Using the first value, one can calculate $\hat{f}(x_1(1)) = \frac{1}{10 \times 1.0690} \sum_{j=1}^{10} \frac{1}{\sqrt{2\pi}} \cdot e^{-\frac{1}{2}\left(\frac{5.6091 - x_1(j)}{1.0690}\right)^2} = 0.1769$. This step is repeated for every point, resulting in the KDE provided in Table 4.5.

Step 2 All these steps were preparations for the optimization algorithm, where repeated calculations of the objective function will be bypassed through the reweighting scheme.

Step 2a For the first iteration, assume that the initial value for the decision variables is $x_1 = 5$ and $x_2 = 5$. For these values, another sample set is generated, as shown in Table 4.6, accounting for the uncertainties described in Eqs. 4.9 and 4.10.

Table 4.5 Base sample kernel density estimates

x_1	$\hat{f}(x_1)$	x_2	$\hat{f}(x_2)$
5.6091	0.1769	0.3573	0.1277
3.7217	0.0932	1.9974	0.2114
6.2927	0.2046	4.2713	0.1602
7.2671	0.2000	3.3062	0.2190
4.1182	0.1110	1.3274	0.2068
7.7831	0.1711	1.5233	0.2117
6.9578	0.2090	1.1575	0.1992
5.4475	0.1691	3.6813	0.2100
8.8302	0.0920	2.9210	0.2152
6.9428	0.2092	3.7507	0.2063

Table 4.6 Sample-optimization iteration 1

Sample no.	$\tilde{x}_1$	$\tilde{x}_2$
1	4.7790	5.7625
2	4.9029	5.5740
3	5.0347	5.9199
4	4.9686	5.8697
5	4.9001	5.9967
6	4.9819	5.1281
7	5.0316	5.4877
8	5.0403	5.4841
9	4.9447	5.7557
10	5.0344	4.7531
Mean	4.9618	5.5731
Std. Dev	0.0836	0.3862

The expected value of Z is estimated using the reweighting approach, given in Steps 2b and 2c.

Step 2b Now, the KDE for the sample ($f(x_i)$) generated around the decision variables has to be calculated. The Gaussian kernel width $h(\tilde{x}_1) = 1.06 \times 0.0837 \times 10^{-0.2} = 5.598 \times 10^{-2}$. Using this value, one can calculate $f(x_1(1)) = \frac{1}{10 \times 5.598 \times 10^{-2}} \sum_{j=1}^{10} \frac{1}{\sqrt{2\pi}} \cdot e^{-\frac{1}{2}\left(\frac{5.609 - \tilde{x}_1(j)}{5.598 \times 10^{-2}}\right)^2} = 5.125 \times 10^{-23}$. Again, this step is repeated for every point of the sample with respect to the base distribution data resulting in the KDE provided in Table 4.7.

Step 2c Using these and the base KDE values, weights are calculated for each sample point j as

$$\omega_j = \frac{f(x_1(j))}{\hat{f}(x_1(j))} \times \frac{f(x_2(j))}{\hat{f}(x_2(j))}, \quad j = 1, ..., N \tag{4.15}$$

Table 4.7 Optimization iteration 1-KDE

No	x_1	$f(x_1)$	x_2	$f(x_2)$
1	5.6091	5.125×10^{-23}	0.3573	0
2	3.7217	0	1.9974	2.989×10^{-26}
3	6.2927	0	4.2713	2.777×10^{-2}
4	7.2671	0	3.3062	2.376×10^{-8}
5	4.1182	3.918×10^{-31}	1.3274	9.958×10^{-40}
6	7.7831	0	1.5233	1.745×10^{-35}
7	6.9578	0	1.1575	1.303×10^{-43}
8	5.4475	5.218×10^{-12}	3.6813	2.826×10^{-5}
9	8.8302	0	2.9210	1.844×10^{-12}
10	6.9428	0	3.7507	8.311×10^{-5}

Table 4.8 Optimization progress at $N = 100$

Iteration	x_1	x_2	$E^{est}[Z]$
0	5.000	5.000	5.958
1	9.610	2.353	9.238
2	7.065	3.814	0.258

In our illustrative example, the only two nonzero weights are $\omega_5 = 1.699 \times 10^{-68}$) and $\omega_8 = 4.152 \times 10^{-15}$. These weights are normalized and multiplied with the output of the base distribution to estimate the objective function value:

$$E^{est}[Z] = \sum_{j}^{N} \overline{\omega_j} \cdot Z(j). \qquad (4.16)$$

For our illustrative example, this reduces to

$$E^{est}[Z] = \overline{\omega_8} \cdot Z(8) = 1.0000 \times 2.5119 = 2.5119, \qquad (4.17)$$

as the normalization eliminates all but one weight. Note that this illustrative example was developed with an unrealistically small sample size. Hence, the accuracy of the estimation technique cannot be judged from this example. Further, due to the inaccuracy of the estimate resulting from the small sample size, we will not present results for Steps 2d and 2e for just 10 samples, but use 100 samples (note that estimated value of expected value of Z is different in Table 4.8, and is different than that of 10 samples). Also note that Steps 2d and 2e basically repeat the procedures in Steps 2a through 2c for a new sample set around a perturbed point, for instance $x_1 + \Delta x_1 = 5 + 0.001 \times 5 = 5.005$.

The results obtained using the BONUS algorithm for optimization converge to the same optimal solution as obtained using a brute force analysis normally used in stochastic NLPs where the objective is calculated for each iteration by calculating the

Fig. 4.5 Nonisothermal
CSTR

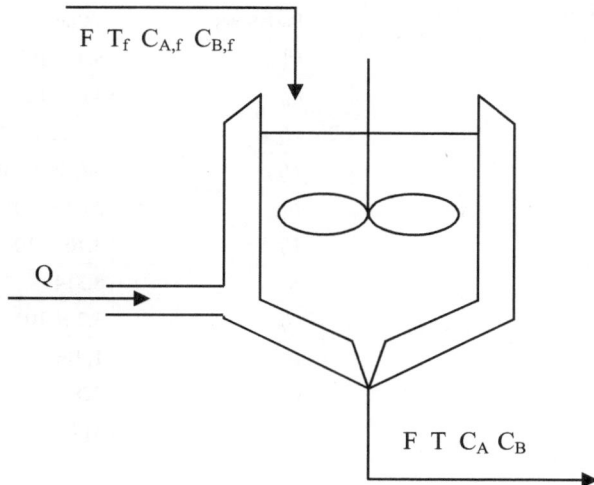

$$F \ T_f \ C_{A,f} \ C_{B,f}$$

$$Q$$

$$F \ T \ C_A \ C_B$$

objective function value for each generated sample point. In this example, BONUS
used only 100 model runs, while the brute force optimization evaluated the model
600 times for the two iterations.

The following example is based on Taguchi's approach to off-line quality control
[55] applied to output of a chemical reactor system.

Example 4.2, Taguchi's Quality Control Problem Consider the following problem of off-line quality control of a continuous stirred tank reactor (CSTR) derived
from [23].

The system to be investigated consists of a first-order sequential reaction,
$A \rightarrow B \rightarrow C$, taking place in a nonisothermal continuous CSTR. The process and the associated variables are illustrated in Fig. 4.5. We are interested in
designing and operating this process such that the rate of production of species B
(R_B) is 60 moles/min. However, as is apparent from the reaction pathway, species
B degrades to species C if the conditions in the CSTR such as the temperature (T)
and heat removal (Q) are conducive. The objective of parameter design is to produce species B at target levels with minimal fluctuations around the target in spite
of continuous variation in the inputs. The inlet concentration of A (C_{A_f}), the inlet
temperature (T_f), the volumetric flow rate (F), and the reactor temperature (T) are
considered prone to continuous variations. The objective of off-line parameter design
is to choose parameter settings for the design variables such that the variation in the
production rate of r_B around the set point is kept at a minimum.

The five design equations that govern the production of species B (and the steady
state values of other variables) in the CSTR are given below. The average residence
time (τ) of each species in the reactor is given as $\tau = V/F$, where V is the reactor
volume and F is the feed flow rate.

Table 4.9 Parameters and their values in CSTR study

Parameter	Value	Units
k_A^0	8.4×10^5	min^{-1}
k_B^0	7.6×10^4	min^{-1}
H_{RA}	-2.12×10^4	J/mol
H_{RB}	-6.36×10^4	J/mol
E_A	3.64×10^4	J/mol
E_B	3.46×10^4	J/mol
R	8.314	J/(mol $\cdot$ K)
C_p	3.2×10^3	J/(kg $\cdot$ K)
ρ	1,180	kg/m^3
C_{B_f}	328	mol/m^3
T	314	K

$$Q = F\rho Cp(T - T_f) + V(r_A H_{RA} + r_B H_{RB}) \qquad (4.18)$$

$$C_A = \frac{C_{A_f}}{1 + k_A^0 e^{\frac{-E_A}{RT}} \tau} \qquad (4.19)$$

$$C_B = \frac{C_{B_f} + k_A^0 e^{\frac{-E_A}{RT}} \tau C_A}{1 + k_B^0 e^{\frac{-E_B}{RT}} \tau} \qquad (4.20)$$

$$-r_A = k_A^0 e^{\frac{-E_A}{RT}} C_A \qquad (4.21)$$

$$-r_B = k_B^0 e^{\frac{-E_B}{RT}} C_B - k_A^0 e^{\frac{-E_A}{RT}} C_A \qquad (4.22)$$

where C_A and C_B are the bulk concentrations of A and B, T is the bulk temperature of the material in the CSTR, subscript f denotes initial feed, and the rate of consumption of A and B are given by $-r_A$ and $-r_B$. These five variables are the state variables of the CSTR and can be estimated for a given set of values for the input variables (C_{A_f}, C_{B_f}, T_f, T, F, and V) and the following physical constants: k_A^0, k_B^0 and E_A, E_B the preexponential Arrhenius constants and activation energies respectively; H_{RA} and H_{RB}, the molar heats of the reactions, which are assumed to be independent of temperature; ρ and Cp the density, and specific heats of the system, which are assumed to be same for all processing streams. Once input variables T and T_f are specified, Eq. 4.18 can be numerically solved to estimate Q, the heat added to or removed from the CSTR. The average residence time can be calculated from the input variables F and V. Subsequently, for a given input concentration for C_{A_f} and C_{B_f}, the bulk CSTR concentrations C_A and C_B can estimated using Eqs. 4.19 and 4.20. The production rates r_A and r_B can now be calculated from Eqs. 4.21 and 4.22. The system parameters are summarized in Table 4.9. Note that this analysis fixes the set-point for both the feed concentration of B, C_{B_f}, and the CSTR temperature T. Both values are also given in Table 4.9.

Table 4.10 Decision variables for optimization

	Lower bound	Upper bound	Initial value	Optimal value
C_{A_f}	3000 mol/m^3	4000 mol/m^3	3118 mol/m^3	3125.1 mol/m^3
T_f	300 K	350 K	314 K	328.93 K
F	0.01 m^3/min	0.1 m^3/min	0.070 m^3/min	0.057 m^3/min
V	0.02 m^3	0.05 m^3	0.0391 m^3	0.0500 m^3

The design objective is to produce 60 mol/min of component B, i.e., $R_B = 60$. The initial nominal set points for the decision variables are provided in Table 4.10. However, the continuous variations in the variables (C_{A_f}, T_f, F, and T) result in continuous variations of the production rate, R_B, which needs to be minimized. Solve this problem using traditional SNLP and BONUS, and compare the results.

Solution The goal is to determine process parameters for a nonisothermal CSTR (Fig. 4.5) that result in minimum variance in product properties when fluctuations are encountered [23]. The mathematical representation for the problem is given as:

$$\min \sigma^2_{R_B} = \int_0^1 (R_B - \overline{R_B})^2 dF \tag{4.23}$$

$$\text{s.t. } \overline{R_B} = \int_0^1 R_B(\theta, x, u) dF \tag{4.24}$$

$$C_A = \frac{C_{A_f}}{1 + k_A^0 \cdot e^{-E_A/RT} \cdot \tau} \tag{4.25}$$

$$C_B = \frac{C_{B_f} + k_A^0 \cdot e^{-E_A/RT} \cdot \tau \cdot C_A}{1 + k_B^0 \cdot e^{-E_B/RT} \cdot \tau} \tag{4.26}$$

$$-r_A = k_A^0 \cdot e^{-E_A/RT} \tag{4.27}$$

$$-r_B = k_B^0 \cdot e^{-E_B/RT} - k_A^0 \cdot e^{-E_A/RT} \tag{4.28}$$

$$Q = F\rho C_p \cdot (T - T_f) + V \cdot (r_A H_{RA} + r_B H_{RB}) \tag{4.29}$$

$$\tau = V/F \tag{4.30}$$

$$R_B = r_B \cdot V \tag{4.31}$$

Uncertain variables are $[C_A, T_f, F, T]$, and the range of uncertainty for these variables is normally distributed with means at $[C_{A_f}, T_f^1, F^1, T^1]$. For the first three uncertain variables, the fluctuations 0.001^{th} fractiles are at $\pm 10\%$. However, for T, several factors can contribute to fluctuations and the level of fluctuation around the reactor temperature T is set at $\pm 30\%$ around T^1. Based on these values, the initial variance at the starting point given in Table 4.10 is determined as $\sigma^2_{R_B,init} = 1034$.

To compare the performance of bypassing the model and using the estimation technique through kernel densities, the model was run first for the case with traditional SNLP. Using this traditional approach, the algorithm converged to the optimal

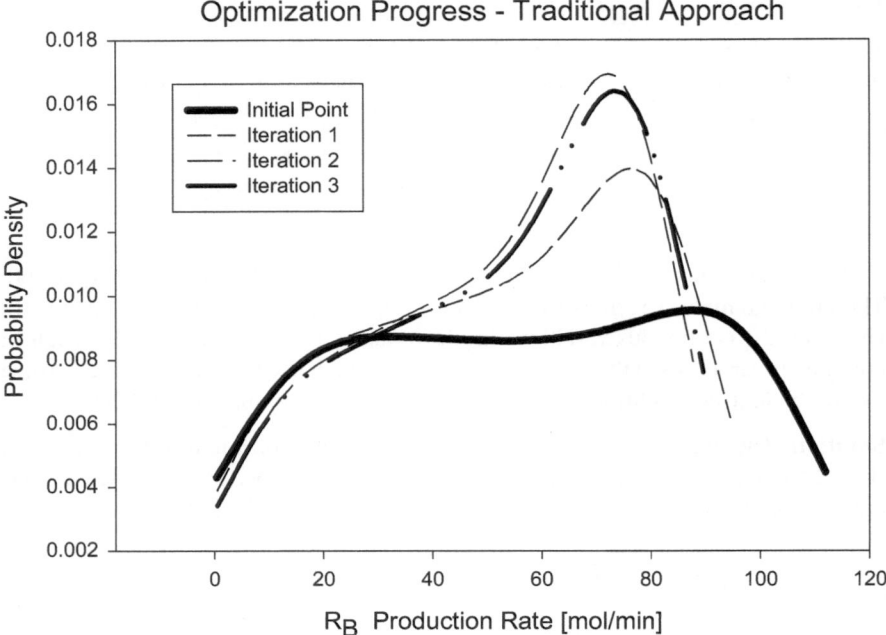

Fig. 4.6 Optimization progress for traditional SNLP approach

solution of $[C_{A_f} = 3124.7 \text{ mol/m}^3, T_f^1 = 350 \ K, F^1 = 0.0557 \text{ m}^3/\text{min}, V = 0.0500 \text{ m}^3]$ after three iterations, for a sample size $N_{samp} = 150$. This reactor design has a variance of $\sigma_{R_B}^2 = 608.16$. Here, the model is run for every sample point during each iteration step. Further, the derivatives used for SQP are estimated by running the model an additional four times for shifted sample sets of each variable. This requires a total of

$$150\frac{\text{model calls}}{\text{derivative calc.}} \cdot (4+1)\frac{\text{derivative calc.}}{\text{iterations}} \cdot 3 \text{ iteration} = 2250 \text{ model calls}$$

Optimization progress is presented in Fig. 4.6 for the traditional approach and in Fig. 4.7 for BONUS. The initial point is shown as the thick line covering variations up to 120 mol/min. As optimization progresses, the probability around the desired rate of $R_B = 60$ increases, as seen in the optimal solution presented as the bold dashed/dotted line.

The analysis for the BONUS algorithm using model bypass converges after five iterations to the same optimum values of decision variables $C_{A_f} = 3125.1 \ mol/\text{m}^3$, $T_f = 328.93K$, $F = 0.057\text{m}^3/\text{min}$, and $V = 0.0500 \text{ m}^3$. This solution shows almost identical behavior to the optimum found using the traditional approach and even has a slightly lower variance of $\sigma_{R_B}^2 = 607.11$. However, the real advantage of using BONUS is that this analysis called the model just 150 times, only for the

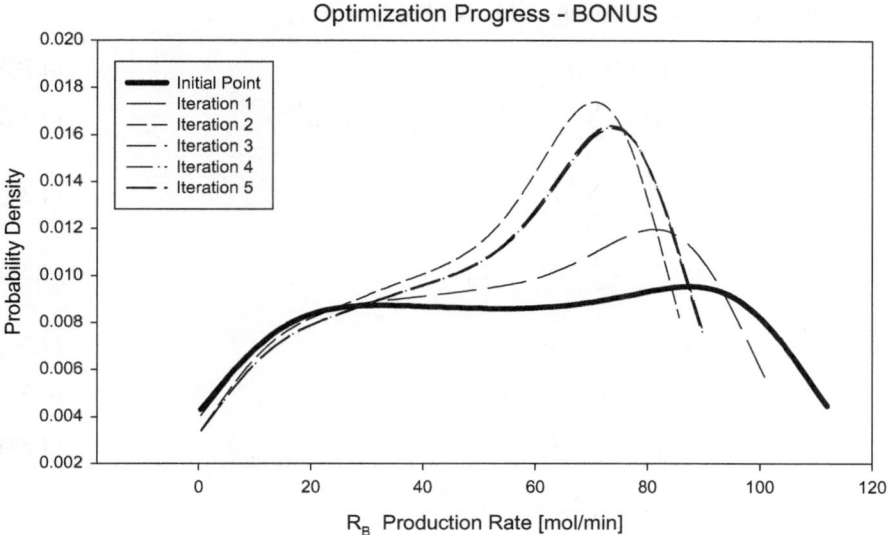

Fig. 4.7 Optimization progress in reducing product variance using BONUS

determination of the initial base distribution $F[R_B^\star]$, in contrast to a total of 2250 model evaluations for the traditional approach.

Capacity expansion for electricity utilities has been an active area of research, having been analyzed using a multitude of methods, including optimization, simulation, and decision analysis [27]. The nature of the problem is inherently uncertain, as it is impossible to determine exact values for future cost levels, the demand for electricity, the development of alternative and more efficient technologies, and many more factors. Hence, the capacity planning example has been analyzed by various researchers in the stochastic programming (SP) community [2] .

Due to the limitations of conventional algorithms for optimization under uncertainty, several assumptions have been made, converting the capacity expansion SP into a linear problem through estimations and approximations in order to solve these problems. Among these simplifications, the load curve, which identifies the probability of electricity demand levels, is generally discretized into linear sections, allowing the use of decomposition techniques that require a finite number of realizations of the uncertain variables [30]. The ability of BONUS to handle nonlinearity allows this problem to be handled without these limitations; this is presented in the following examples.

Example 4.3 Capacity Expansion for Electric Utilities The mathematical representation of the problem is given below. The objective is to minimize the expected cost of capacity expansion subject to uncertain demands and cost factors, while ensuring that no shortages are present. Note that the objective function 4.32 is the expected value for the total cost calculated for $n = 1, ..., N_{samp}$ samples. In the formulation given below, capital nomenclature is used for decision variables, while the uncertain variables are indicated through a tilde symbol.

$$\min E[cost] \tag{4.32}$$

$$\text{s.t. } cost = \sum_t cost_t^{op} + cost_t^{cap} + cost_t^{buy} \tag{4.33}$$

$$cost_t^{op} = \sum_i P_t^i \cdot \widetilde{oc}_t^i \tag{4.34}$$

$$cost_t^{cap} = \sum_i \alpha^i \cdot (AC_t^i)^{\beta^i} \tag{4.35}$$

$$cost_t^{buy} = \widetilde{\kappa}_t \cdot (\widetilde{d}_t - tp_t)^{\gamma} \tag{4.36}$$

$$c_t^i = c_{t-1}^i + AC_t^i \tag{4.37}$$

$$tp_t = \sum_i P_t^i \tag{4.38}$$

$$P_t^i \leq c_t^i \tag{4.39}$$

$$i \in \text{Technology}_1, \text{Technology}_2, ..., \text{Technology}_I \tag{4.40}$$

$$t \in \text{Period}_1, \text{Period}_2, ..., \text{Period}_T \tag{4.41}$$

Equation 4.33 sums up the respective costs for operation, capacity expansion, and the option to purchase electricity for meeting demand in case the total available capacity is below demand. The operating costs are calculated using Eq. 4.34, where oc_t^i is a cost parameter for electricity generation of technology i in time period t, and P_t^i are decision variables determining how much electricity should be produced using technology/power plant i at time t.

Equation 4.35 determines the cost of capacity expansion. Traditional models use a linear relationship between the cost of expansion $Cost_t^{cap}$ and the added capacity ac_t^i. Use the data and models from the Integrated Environmental Control Model (ICEM), a computational tool developed for the Department of Energy. This will provide the power law model for more accurate cost estimation (Eq. 4.35). In this formula, α_i is a proportionality factor for capacity expansion, while β_i provides the exponential factor that allows capital expansion cost to follow economies of scale.

Another nonlinear expression, Eq. 4.36, will be used to determine the cost of electricity purchased, $Cost^{buy}$, when demand d_t exceeds capacity. The power factor γ must be greater than 1 to ensure that relying on external sources is not used as the sole option when increase in demand is expected. This is accomplished as $Cost^{buy}$ increases exponentially when capacity is significantly below possible demand levels. The primary goal of this approach is to account for the common market practice of purchasing electricity in a deregulated environment when demand reaches peak levels, surpassing available capacity in a given location.

Finally, use Eq. 4.37 to calculate the available capacity at each time step following expansion, Eq. 4.38 calculates the total electricty produced, tp_t, and Eq. 4.39 ensures that no power plant can produce more energy than its installed capacity.

Table 4.11 Uncertain variables in capacity expansion case

Parameter	Lower bound	Upper bound
Demand Period I	400 MWh	500 MWh
Demand growth rate Period II	0.75	1.50
Tech. I Gen. cost increase for P-II	0.95	1.12
Tech. II Gen. cost for Period I	0.17 k$/MW h	0.37 k$/MW h
Tech. II Gen. cost for Period II	0.17 k$/MW h	0.50 k$/MW h

Table 4.12 Constants for capacity expansion case

Parameter	Value
Initial capacity Tech. I	250 MW
Initial capacity Tech. II	150 MW
β^1	1.25 k$/MW$^{0.7472}$
β^2	0.95 k$/MW$^{0.7856}$
α^1	0.7472
α^2	0.7856
γ	1.75
oc_1^1	0.25 k$/MWh

In this problem, Technology I is selected as a Cyclone type coal power plant, while Technology II is a Tangential plant. Again, data for these technologies can be obtained using IECM [41] model.

There are five uncertain variables (Table 4.11) and eight decision variables that determine capacity expansion and electricity generation for each technology at each time step.

Here, demand growth rate for Period II implies that the total demand in Period I is multiplied by a normally distributed uncertain factor varying between 0.75 and 1.50, while the unit cost of electricity generated through Technology I can vary between -5 and $+12\%$ for the second period. Table 4.12 provides the constants and initial values used for this case study.

Finally, the preexponential factor for the cost of purchasing electricity, κ_t can be determined as the greater value between the two per unit electricity generation costs for the different technologies, oc_t^1 and oc_t^2.

Solve this problem using traditional SNLP and BONUS.

Solution Starting from a system with initial annualized cost of the capacity expansion at $760.9K, the system is optimized both via BONUS, as well as exhaustive model runs for derivative estimation through objective function value calculation. The conventional approach converges after five iterations, requiring a total of

$$100\frac{\text{model calls}}{\text{derivative calc.}} \cdot (8+1)\frac{\text{derivative calc.}}{\text{iterations}} \cdot (5)\,\text{iteration} = 4,500\,\text{model calls}$$

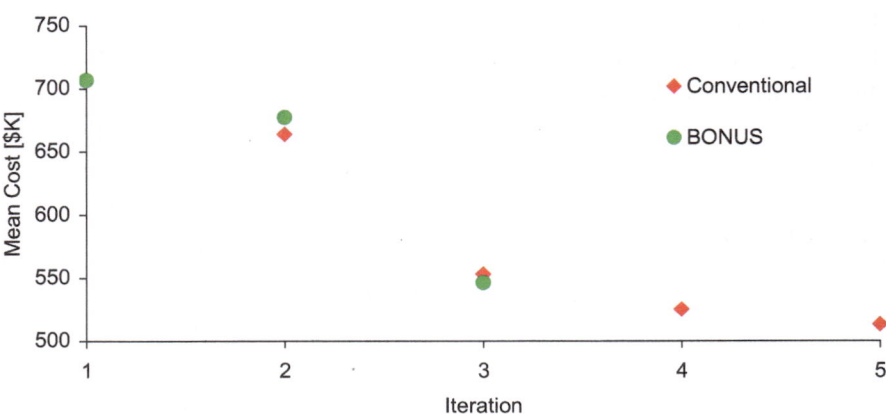

Fig. 4.8 Comparison of optimization progress

Table 4.13 Decision variables in capacity expansion case

Variable	Initial value (MW)	Optimal Value (MW)
Capacity addition Tech I Period I	100	93
Capacity addition Tech I Period II	200	197
Capacity addition Tech II Period I	100	154
Capacity addition Tech II Period II	200	197
Electricity production Tech I Period I	250 h	257 h
Electricity production Tech I Period II	250 h	247 h
Electricity production Tech II Period I	250 h	291 h
Electricity production Tech II Period II	250 h	330 h

compared to only 100 models and just three iterations run for the BONUS algorithm
(Fig. 4.8). Table 4.13 presents the decision variables and their optimal values found
by BONUS.

$$\epsilon_{BONUS} = \frac{(\% \text{ Mean reduction})_{\text{BONUS}}}{(\% \text{ Mean reduction})_{\text{Modelruns}}} = 0.867$$

4.4 Summary

In this chapter, we have introduced the BONUS based on the reweighting approach for
estimating derivative information needed during optimization of nonlinear stochastic problems. The technique relies on KDE of a base distribution and the sample

space encountered during optimization. Two real world case studies; (1) an off-line quality control problem from chemical engineering, and (2) the electricity expansion problem from operations research literature, illustrates efficiency of the technique in determining derivatives, and hence the search directions during optimization loop. Further, by selection of efficient sampling techniques like HSS allows for significant computational improvement, as the repetitive nature of model evaluations is avoided by using the reweighting scheme. The BONUS algorithm is very useful for solving large-scale real-world problems of significance (e.g., for black-box models) is illustrated in the following three chapters.

Notations

AC_t^i	capacity of technology i in time period t
C_i	concentration of component i in mol/m^3
C_{i_f}	inlet concentration of component i in mol/m^3
$cost_t^{op}$	operating cost
$cost_t^{cap}$	capacity expansion cost
$cost_t^{buy}$	cost of buying electricity
d_t	electricity demand
D	number of decision variables
E	expected value function
f	probability density function
F	volumetric flowrate, m^3/min
oc_t^i	a cost parameter for electricity generation of technology i in time period t
P	probability function
P_t^i	decision variables determining how much electricity should be produced using technology/power plant i at time t
Q	heat removal J
N_{samp}	number of samples
R_B	rate of production of B
$\overline{R_B}$	average rate of production of B
S	number of uncertain variables
T	temperature, ^0K
T_f	inlet temperature, ^0K
u	uncertain and decision variable from input distributions
$u*$	base uncertain and decision variable from input distributions
v	uncertain variable
V	reactor volume, m^3
Z	output variable

Greek letters

α_i	a proportionality factor for capacity expansion

β_i	an exponential factor that allows capital expansion cost to follow economies of scale
γ	power factor
θ	decision variable
μ	mean
σ	standard deviation
τ	residence time, min
ω	weighting function

Chapter 5
Water Management Under Weather Uncertainty

5.1 Introduction

Water scarcity and the cost of treating and recycling waste water both represent constraints in operating coal-fired power plants. As the capacity of thermoelectric power generation increases in the USA (the Energy Information Administration estimates that thermoelectric power generation will grow 22 % by 2030), so does the importance of managing the water used in these plants. In a clean coal-fired power plant, water is consumed in makeups (water added to a closed cycle due to evaporation or product loss), in blowdowns (water added during the cooling cycle due to liquid removal), and in the generation process itself. The amount of water consumed varies with two ambient weather factors: the dry-bulb temperature (temperature as measured by a thermometer shielded from moisture) and the humidity of the outside air, both of which are subject to significant uncertainty, and vary with the season and geographical region. It is, therefore, critical to determine the optimal operating conditions for these plants, so as to minimize water consumption subject to stochastic weather conditions. In this chapter, it is demonstrated how the BONUS method can substantially simplify this problem for a pulverized coal (PC) power plant. This chapter is based on work by Salazar et al. [44, 45].

5.2 The Pulverized Coal Power Plant

The specific PC power plant model referenced herein is based on Case 11 in the DOE/NETL's report on the cost and performance of fossil energy plants (NETL, 2010). The model is a supercritical steady-state flowsheet without carbon capture designed to generate 548 MW of electricity.

A pulverized coal power plant generates electricity in four steps (as shown in Fig. 5.1):

© Urmila Diwekar, Amy David 2015   57
U. Diwekar, A. David, *BONUS Algorithm for Large Scale Stochastic Nonlinear
Programming Problems,* SpringerBriefs in Optimization, DOI 10.1007/978-1-4939-2282-6_5

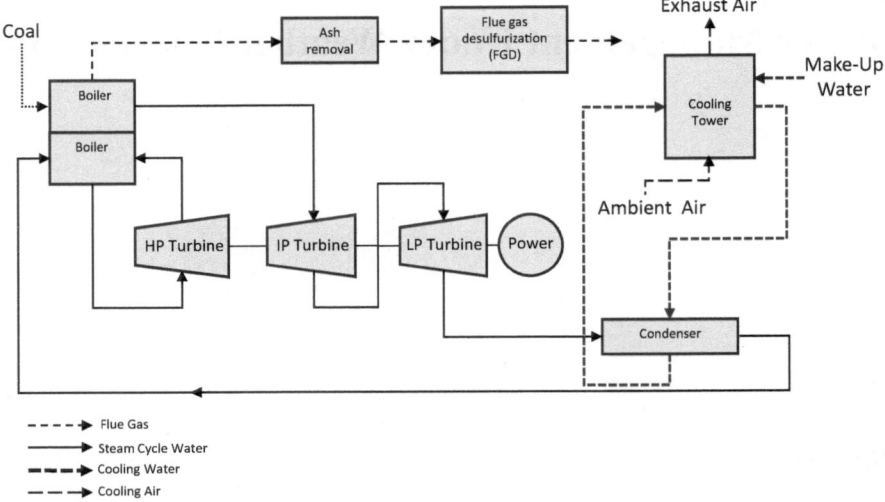

Fig. 5.1 Schematic of PC power plant

1. Powdered coal is fed into the boiler's combustion chamber, where the combustion of coal produces steam.
2. The steam is fed into a series of high (HP), intermediate (IP), and low pressure (LP) turbines, causing the turbines to rotate at high speed.
3. The steam is cooled to condensation in a cooling tower.
4. The condensation is preheated with steam extracted from the turbines and returned to the boiler.

Subsequent to the generation process, the gaseous waste is passed through a flue gas desulfurization (FGD) unit prior to its atmospheric release. The FGD unit removes sulfur dioxide (SO_2) by combining the SO_2 with limestone slurry and oxygen to produce calcium sulfate (gypsum). The gypsum is then separated from the water, which is then recycled, yet a large amount of makeup water is still required to replace that lost in the desulfurization process.

Therefore, considering both the power generation process and the FGD unit, the areas of water consumption are:

- Water lost to evaporation in the cooling tower in the third step of the power generation process (EL)
- Water lost to drift in the cooling tower during the power generation process (D)
- Water used for "blowdown" in the cooling tower (B)
- Water used in the FGD for preparation of the limestone slurry and makeup (F)

The process performance parameters (generation, efficiency, emission, and water consumption) are simulated in Aspen Plus®, a chemical process modeling system from Aspen Technologies Inc. Aspen Plus® is commonly used for modeling power plants because of its capabilities in representing multiphase streams and handling complex substances such as coal [10]. In this case, Aspen Plus® takes the process design and operational parameters as its inputs and outputs the process performance parameters. Because the optimization techniques detailed in this chapter seek to find the optimal inputs to minimize or maximize performance parameters, the robustness of the Aspen Plus® model is critical to the reliability of the optimization results.

Within this boiler/turbine/condenser cycle, water is lost to evaporation associated with the quantity of heat rejected at the cooling tower. To estimate the evaporation rate, an equilibrium-based model for the cooling tower (based on a scheme proposed by [15] is implemented in Aspen Plus® as what is known as a unit-operation-based model. Each "unit block" is a simulation unit that allows the user to define calculations not native to Aspen Plus®. Three unit blocks are used in this model, two flash separators and one heat exchanger. The first flash separator is used to determine the wet bulb temperature from the dry-bulb temperature and the humidity, while the second simulates the cooling tower itself. Specific details on the internal calculations of each of these blocks can be found in [44]. Using the design specifications and the calculator blocks, Aspen Plus® is able to determine the cold water temperature, circulating water flow rate, and air flow rate for a constant volume forced drift cooling tower, and thus calculate the water usage due to evaporation.

In addition to the evaporation losses, water is consumed due to both drift and blowdown. "Drift" refers to the water caught in the air leaving the top of the tower, estimated as 0.02 % of the water circulating through the cooling tower. It differs from evaporation in that the drift water remains in liquid form, while evaporation is the water that has been converted to steam, and thus the two types of loss are considered separately in calculations, though both contribute to the necessity of "makeups" and water that must be added to a closed cycle to compensate for losses.

Blowdown, by contrast, is water added to appropriately dilute corrosive substances. The amount of water consumed as blowdown (B) is estimated as

$$B = \frac{EL - (C - 1)D}{C - 1},$$

where C is the number of concentrating cycles (assumed to be 4), EL is the evaporation loss, and D is the drift loss.

The evaporation losses (ELs) are dependent on both temperature and humidity at the plant location, and because the drift and blowdown are calculated from the evaporation, all three are dependent upon operating conditions and weather factors. The water used in the FGD unit is dependent upon the conditions of the flue gas coming out of the boiler, which is, in turn, dependent on both the operating conditions and weather factors as well. The goal is, therefore, to determine the operating conditions that minimize the expected value of water consumption subject to the uncertainty in the ambient weather.

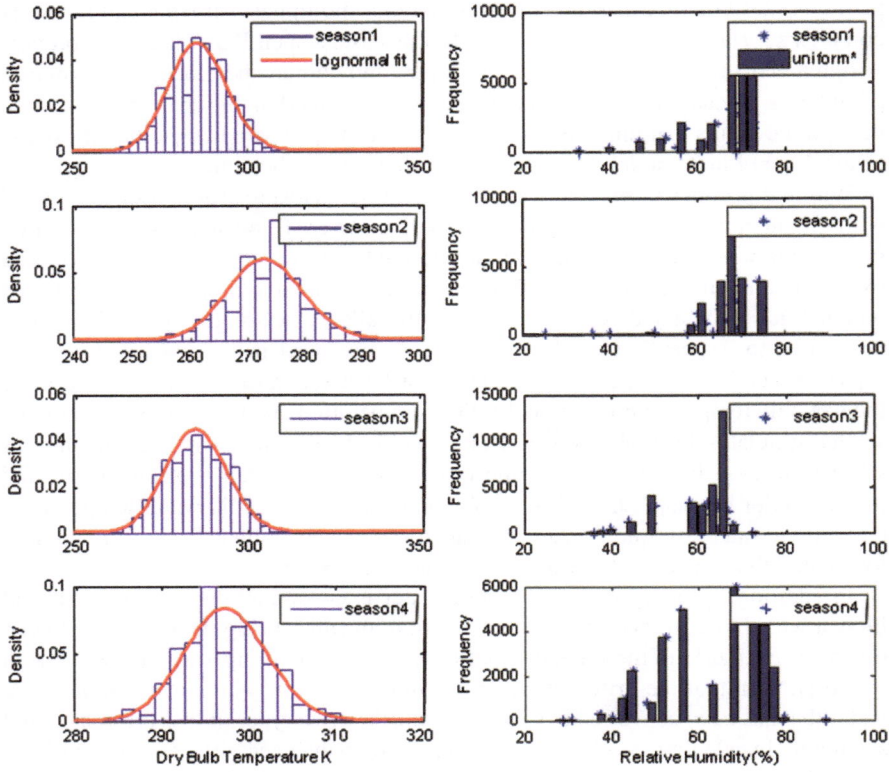

Fig. 5.2 Probability density functions of four seasons (fall to winter from *top*) for dry-bulb temperature and relative humidity in eight US Midwestern cities

5.3   Parameter Uncertainty

Dry-bulb temperature and dew point data for US urban centers is available at the DOE's energy efficiency and renewable energy program (EERE) website (http://www1.eere.energy.gov). Two years of dry-bulb temperature and dew point data, spanning September 2005 through August 2007, was averaged for eight major Midwestern US cities (Chicago, Detroit, Indianapolis, Minneapolis, St. Louis, Des Moines, Kansas City, and Cincinnati). This dry-bulb temperature data was organized into bins and histograms were generated for each season. For each bin, the corresponding dew point values were averaged to calculate values for relative humidity, and thus histograms could be generated for this parameter as well. The frequency distributions for dry-bulb temperature and humidity are shown in Fig. 5.2.

   The probability distribution of the dry-bulb temperature was fitted to a lognormal distribution and the relative humidity was fitted to a modified uniform distribution (in which the range is divided into different intervals within which all values have an equal likelihood of occurrence). A distribution was created for each of the four seasons, using the eight-city average. It is notable that both parameters have larger variations during fall and spring than during winter and summer. This variability will, in turn, result in a higher variability for water consumption in these seasons.

Once the probability distribution functions are estimated, stochastic simulation and optimization are carried out as follows:

1. Efficiently sample the probability functions of dry-bulb temperature and relative humidity to generate a set of scenarios that accurately represent potential realizations of uncertainty.
2. Propagate uncertainty by executing the whole plant simulation for every scenario and record the water consumption in each.
3. Analyze the resulting distribution of water consumption and choose either the expected value or standard deviation (the first or second moment) as the objective function for stochastic optimization.

5.4   Problem Formulation

The objective of this problem is to minimize the expected value (**E**) of water consumption in the PC power plant, expressed as the sum of the evaporation losses (EL), drift losses (D), blowdown losses (B), and FGD consumption (F). The deterministic input parameters, x, are the design and operational conditions of the units, and the stochastic input parameters, u, are the uncertain weather conditions.

Further, let Q represent the total heat rejected by the cycle and P represent the amount of power generated, both dependent on x and u. Similarly, evaporation loss (EL) depends on the total heat rejected by the cycle (Q), drift (D) depends on (EL), and blowdown B depends on both D and EL, while water consumption in the FGD unit also depends on x and u. Thus, the following equations may be introduced: $Q = f(x, u)$, $P = h(x, u)$, $EL = g_1(x, u)$, $D = g_2(x, u)$, $B = g_3(x, u)$, and $F = g_4(x, u)$, where the set of functions, f, h, and g_i represent "black-box" calculations in Aspen Plus®, i.e., the exact nature of the functions is proprietary, and knowledge of such is unimportant to the problem formulation and solution.

The problem can therefore be represented as

$$\min_{x_d} \mathbf{E}(g(x_d, u))$$

$$s.t.$$

$$h(x_d, u) - P^* = 0$$

$$ME(x_d, u) = 0,$$

where P^* is the fixed generation of electricity and ME is the material and energy balance of the operating unit. Thus, the first constraint ensures that the required amount of electricity is generated and the second ensures that mass and energy balances are respected. Because the functions f, ME, and h are highly nonlinear, this problem is extremely computationally complex, and a method such as the BONUS algorithm is needed to make the problem tractable.

5.5   Selection of Decision Variables

Of the assigned parameters that potentially influence water consumption, as detailed in the NETL report on cost and performance baselines for fossil energy plants [36], nine were selected as potential decision variables:

1. Boiler temperature: The temperature of the unit that burns coal particulate to initially heat water into steam.
2. Air excess: The amount of air present in the combustion chamber in excess of the theoretical minimum, used to ensure all coal particulate is exposed to sufficient air.
3. Reheater temperature: The temperature of the unit that reheats the steam between the high pressure turbine and the medium pressure turbine.
4. FGD efficiency: A measure of the rate of SO_2 removal in the FGD, influenced by the design parameters of the unit, such as surface area and absorption material.
5. O_2/SO_2 ratio: the ratio of oxygen to sulfur dioxide in the FGD unit.
6. $CaCO_3/SO_2$ ratio: The ratio of calcium carbonate to sulfur dioxide in the FGD unit.
7. Water content of FGD slurry: The amount of water added to limestone to produce the FGD slurry.
8. Pressure drop at high-pressure condenser 1: The difference in pressure between the steam exhausted from the turbines and that leaving the first condenser.
9. Pressure drop at high-pressure condenser 2: The difference in pressure between the steam exhausted from the first condenser and that leaving the second condenser.

These nine were selected based on the feasibility of implementation, i.e., these variables are most easily controlled in a practical setting. For example, the water content in the FGD slurry is easily modified through a change in operational policy.

To determine which of the nine operating parameters should be used as decision variables, a stochastic simulation was run in Aspen Plus®, as described in [8]. The ranges of potential decision variables were sampled from uniform distributions, and the model was run for each of the generated combinations, each time producing an output result in the form of a water consumption value. Partial rank correlation coefficients (PRCC) were then calculated as shown in Table 5.1. PRCC are a measure of the relationship between the output and input variables for a nonlinear function; thus the absolute value of the PRCC indicates the influence of the deterministic variable on water consumption. Five variables, air excess, reheater temperature, water content of FGD slurry, pressure drop at high-pressure turbine 1, and pressure drop at high-pressure turbine 2, were found to have the greatest impact on water consumption, and were therefore chosen as the decision variables for the stochastic optimization problem.

Table 5.1 Partial rank correlation coefficients (PRCC) for relationship between potential decision variables and water consumption

Potential decision variable	Partial ranked correlation coefficient
Air excess	0.256642
Reheater temperature	0.228009
FGD efficiency	−0.125901
Boiler temperature	−0.018852
O_2/SO_2 ratio	−0.021453
$CaCO_3/SO_2$ ratio	−0.032031
Water content of FGD slurry	0.191058
Pressure drop at high-pressure turbine 1	−0.294448
Pressure drop at high-pressure turbine 2	−0.266594

5.6 Implementation of BONUS Algorithm

The BONUS algorithm was applied to this problem as follows:

1. A set of 600 scenarios based on the probability distributions of the uncertain inputs and uniform distributions for the decision variable was generated using a Hammersley sequence sampling technique.
2. The model was run for the scenarios generated in step 1 to calculate the value of the objective function and constraints for each set of input values, and a probability distribution was estimated using kernel density estimation for each.
3. The nonlinear optimizer (based on the sequential quadratic programming (SQP) algorithm) was initialized by selecting starting values for each decision variable.
4. The value of the objective function was estimated by first assuming a narrow normal distribution centered at the value chosen in Step 3, and then using this normal distribution, along with the initial uniform distribution of the decision variables and the corresponding outputs found in Step 2 used to calculate new values of the probability density functions according to the following formula:

$$p(u) = \frac{\frac{f(u)}{f(u^*)}}{\sum_{j=1}^{N_{samp}} \frac{f(u_j)}{\widehat{f(u_j^*)}}} p(u_j^*), \tag{5.1}$$

where N_{samp} is the number of samples taken in Step 1, $p(u_j^*)$ is the probability density function for the output distribution corresponding to the initial uniform input distribution, $f(u)$, and $\widehat{f(u)}$ is the probability density function of the updated input distribution. The latter is given by

$$\widehat{f(u)} = \frac{1}{N_{samp}h} \sum_{j=1}^{N_{samp}} \frac{1}{\sqrt{2\pi}} e^{\frac{1}{2}} \left(\frac{u - u_j}{h}\right)^2, \tag{5.2}$$

where h is the variance of the data set.

Table 5.2 Minimization of average water consumption under uncertain air conditions for a 548 MW PC power plant located in the Midwestern US for four different seasons. (Water consumption estimates are reported in millions of pounds per hour)

Season	Optimal values of decision variables				
	Air excess %	RH temperature °F	FGD limestone fraction	HP1 pressure ratio	HP2 pressure ratio
Fall	38.925	1160.8	0.31457	0.49 U	0.61 L
Summer	48.947	1174	0.42262	0.49 U	0.66 L
Spring	35.5	1096.5	0.22185	0.36 L	0.61 L
Winter	19.039	1141.9	0.30077	0.49 U	0.79 U 2
	Base case values of decision variables				
All	20	1157	0.3	0.365	0.637
Season	Bonus estimate optimal objective function	Stochastic simulation estimate optimal objective function	Stochastic simulation base case objective function	Savings %	
Fall	2.433	2.567	2.742	6.4	
Summer	2.622	2.702	3.194	15.4	
Spring	2.421	2.595	2.698	3.8	
Winter	2.331	2.384	2.463	3.2	

5. The decision variables were perturbed, and new estimates of the objective function and its derivative were calculated using the reweighting scheme.
6. Steps 3–5 were repeated until Kuhn–Tucker conditions were reached.

5.7 Results

The nonconvexity of the objective function required the SQP algorithm to be run for a variety of initial values for both reheater temperature and air excess. The nonlinear optimizer was run 61 times, and each nonlinear optimization took between 2 and 20 iterations, for a total of 519 iterations for each of the four seasons. Using the BONUS algorithm, results were derived in 4800 model evaluations per season. By contrast, a traditional framework for this stochastic optimization problem would have instead required at least 120 scenarios in the stochastic loop, for a minimum of 373,680 evaluations, nearly 78 times as many. The BONUS algorithm therefore saved 98.7 % of the computational time required to solve the problem.

Table 5.2 gives the optimal values of the decision variables and water usage estimations at (1) the optimal point using the BONUS algorithm, (2) the optimal point with a rigorous stochastic simulation, and (3) the base case with a rigorous stochastic simulation for each of the four seasons. The water consumption estimations are given as the expected values of the probability distributions approximated with BONUS or calculated via stochastic simulation. The savings in average water consumption at the optimal point, as compared to the base case, range from a low of 3.2 % in winter to 15.4 % in summer.

It is intuitive that the water consumption savings are greater in warmer seasons. In the relatively warm fall and summer seasons, the turbine pressure ratios 1 and 2 are pushed to their upper and lower limits, respectively (these two turbines define the feed-water temperature entering the boiler and the pressure at which steam is reheated). Operating the turbines at the limits of their pressure ratios, along with a higher reheater temperature, increases their thermodynamic generation capacity (work per mass of steam), reducing both required fuel and the steam flow rate. At the same time, these operating parameters have little effect on the heat rejection rate. By combining a reduction in the steam flow rate with a steady heat removal rate, water consumption is minimized in the warmer seasons, in which the cooling tower is least efficient. By contrast, in the cooler seasons (spring and winter), the cooling tower operates more efficiently and allows for reduced water consumption even when a large amount of heat must be removed at the condenser.

5.8   Summary

The BONUS algorithm can be used to efficiently optimize water consumption in a PC power plant. Uncertainty in air temperature and humidity affect the amount of water lost to evaporation, drift, blowdown, and makeup. Reheater temperature, air excess to the boiler, FGD slurry preparation water ratio, and pressure drops at the two high-pressure turbines are all operational variables that may be manipulated to minimize water consumption, but the highly nonlinear nature of the objective function and the power and mass balance constraints make this problem extremely computationally intensive under a traditional stochastic optimization framework. The BONUS algorithm reduces this computational intensity by 98.7 %, and shows that reductions of 3.2–15.4 % are possible, depending on the season, for a 548-MW plant in the Midwestern US.

Notations

B	blowdown losses
D	drift losses
EL	evaporation losses (EL)
f, h, and g_i	constraints using black-box models in Aspen Plus®
$f(u)$	initial uniform input distribution

$\widehat{f(u)}$	probability density function of the updated input distribution
F	FGD consumption
ME	material and energy balance of the operating unit
N_{samp}	number of samples taken
P	amount of power generated
P^*	fixed generation of electricity
$p(u_j^*)$	probability density function for the output distribution corresponding to the initial uniform input distribution
Q	total heat rejected by the cycle
x	set of deterministic input parameters (design and operational conditions)
u	set of uncertain weather conditions

Chapter 6
Real-Time Optimization for Water Management

6.1 Introduction

As discussed in the previous chapter, water consumption represents a critical resource in thermoelectric power generation, and can be challenging to manage due to its dependence on ambient weather conditions. Because weather conditions are both constantly changing and uncertain, a stochastic framework for real-time optimization of power generation is advantageous over a deterministic framework in maximizing power output. In this chapter, it is shown that significant cost savings can be achieved if optimization is done on an hourly basis, and the plant set points are changed accordingly. However, as discussed in the previous chapter, the relationships among plant inputs and outputs are highly nonlinear, and this fact, along with the complexity of both plant operations and weather conditions make the problem extremely computationally intensive. Under a conventional stochastic framework, therefore, an hourly optimization would require a prohibitive level of computing resources. The BONUS reweighting scheme is therefore crucial in devising a tractable approach to real-time optimization. This chapter is derived from Salazar et al. [46].

6.2 Power Plant Operations

Recall from Chap. 6 that in a PC power plant with a wet recirculating cooling system, fuel is burned to generate steam, and the steam, in turn, rotates a turbine, generating electricity [36]. The steam coming out of the turbines must then be condensed in a cooling system before being recirculated to the power plant, as shown in Fig. 6.1. The thermal load at time t, denoted $Q_p(t)$, gives a measure of the thermal energy contained in the steam, and therefore an indication of how much energy must be removed in the cooling process. This thermal load is dependent upon the total amount of power produced by the plant, $P(t)$. These two output variables, $P(t)$ and $Q_p(t)$, comprise

© Urmila Diwekar, Amy David 2015   67
U. Diwekar, A. David, *BONUS Algorithm for Large Scale Stochastic Nonlinear Programming Problems*, SpringerBriefs in Optimization, DOI 10.1007/978-1-4939-2282-6_6

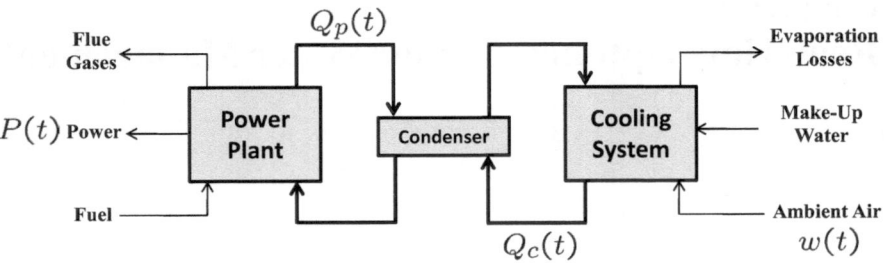

Fig. 6.1 Schematic representation of the interface between the generation and cooling systems

the set of system state variables, $x_p(t)$, and the set of input variables, representing operational parameters such as fuel flow rate, flow of compounds, etc., are denoted $u_p(t)$. Assuming the plant is at steady-state, both inputs and outputs are independent of time; thus x_p and u_p are sufficient representations.

Based on conservation of both mass and energy, an abstract form of the plant model can be written:

$$0 = f_p(x_p, u_p). \tag{6.1}$$

Focusing only on the cooling system, the input parameters can be denoted u_c and the system state variables as x_c. $Q_c(u_c, w)$ is defined as the cooling system capacity, one of the system states, dependent on weather conditions, w, as well as the physical features of the cooling system and u_c. As with the full power plant, the mass and energy balances can be used to write a cooling system model of the form

$$0 = f_c(x_c, u_c, w). \tag{6.2}$$

At steady-state operation, the cooling system capacity should match the capacity required by the power plant, i.e., the values of the operating parameters u_c should be chosen so that the cooling system can exactly meet the needs of the power generation process. While it is obvious that too little capacity would impede power generation, excess capacity is also problematic in that it consumes additional resources such as water, while providing no benefit. This condition is given as

$$0 = g(x_p, x_c) \tag{6.3}$$

The feasible operating region for the cooling system is given by the set of equality constraints,

$$0 = h_c(x_c, u_c, w), \tag{6.4}$$

which depend on the weather conditions and design specifications. This set of constraints can also include bounds on states and decision variables for the cooling system.

The feasible operating region for the power plant in its entirety is similarly given by

$$0 = h_p(x_p, u_p, w), \tag{6.5}$$

a set that includes bounds on states and decision variables for the plant.

Thus, the entire model coupling the power plant and the cooling system can be represented as

$$0 = f(x, u, w), \tag{6.6}$$

$$0 \le h(x, u, w). \tag{6.7}$$

The maximum cooling capacity under *ideal* weather conditions is denoted Q_c^{max}. This maximum capacity is constrained by physical limitations of the system such as size, flooding conditions, pumping capacity, and environmental conditions. These limitations primarily manifest as water constraints, limitations on the amount of fresh water that the cooling system can utilize to provide cooling capacity. For any other *nonideal* weather conditions, the condition $Q_c(u_c, w) \le Q_c^{max}$ holds.

Assume that the maximum power output under design specifications (nominal capacity) is given by P^{max}, with corresponding cooling demand Q_p^{max}. If the available cooling capacity at current conditions is larger than the maximum cooling demand ($Q_c(u_c, w) \ge Q_p^{max}$), power output is constrained only by plant-side design specifications (e.g., furnace capacity), and is equal to the nominal capacity, P^{max}. If, however, the available cooling capacity is smaller than the maximum cooling demand ($Q_c(u_c, w) \le Q_p^{max}$), then the actual maximum amount of power that can be generated is less than the nominal capacity and a function of cooling ability. Let this be represented as $P^{max}(Q_c(u_c, w))$, and note that the power generated must obey the constraint $P \le P^{max}(Q_c(u_c, w))$, because the cooling system must obey the constraint $Q \le Q_p^{max}(Q_c(u_c, w))$, i.e., the cooling system is constrained by the prevailing weather conditions, and the power plant is, in turn, constrained by the cooling system.

It is assumed that the main operational objective in the power plant is to maximize power output at prevailing weather conditions. To achieve this objective, the set-points represented by the operational parameters, u, may be adjusted. For known weather conditions, the optimal set-points can be determined by solving a nonlinear optimization problem of the form

$$\max_u \ J(x, u, w) \tag{6.8}$$

$$s.t. \ 0 = f(x, u, w) \tag{6.9}$$

$$0 \le h(x, u, w), \tag{6.10}$$

where the function $J(x, u, w)$ denotes the plant power output.

6.3   Formulation of the Stochastic Problem

Because, in actuality, the weather conditions are both changing and uncertain, the deterministic problem is insufficient, and a stochastic optimization problem is instead required. The objective of the stochastic problem is to find set-points, u_{HN} for the current time t, that maximize the expected value of the plant capacity. The weather conditions w are assumed to be random variables. The subscript HN indicates the *here-and-now* nature of the solution u_{HN}, that is, the set points are chosen before the actual values of the weather conditions are known. Thus, a single set of set-points is sought, to be implemented under all weather scenarios. This is formulated as

$$\max_{u}\ \mathbb{E}_{w}[J(x(w), u, w)] \tag{6.11}$$

$$s.t.\ \ 0 = f(x(w), u, w), w \in \Omega \tag{6.12}$$

$$0 \le h(x(w), u, w), w \in \Omega, \tag{6.13}$$

where $\mathbb{E}_{w}[\cdot]$ denotes the expected value of the objective function $J(\cdot)$ with respect to w (the expected output power of the plant) and Ω is the set of values over which w has support. The optimal value of this problem is denoted $V(u_{HN}) := \mathbb{E}_{w}[J(x(w), u_{HN}, w)]$. Because the model is assumed to be in steady-state at each time instant, the problems at different time instants are decoupled. Consequently, the stochastic problem can be solved over a horizon $t, ..., t + T$ to determine the expected values $V(u_{HN}(\tau)), \tau = t, ..., t + T$.

6.4   Solution Approach

The stochastic optimization problem can be written as

$$min_{u \in U} V(u) := \mathbb{E}_{w}[\phi(u, w)]. \tag{6.14}$$

In this formulation, U is the box set $U := \{u | \underline{u} \le u \le \overline{u}\}$, where $\underline{u}$ and $\overline{u}$ are the upper and lower bounds, respectively, for the decision variables and uncertain variables together. The function $\phi(u, w)$ results from evaluating $J(x(w), u, w)$ at a given $x(w), w, u$ that solves the model $f(x(w), u, w) = 0$. In other words, it is assumed that inequality constraints exist only in u and the implicit solution of the black box model $f(x, u, w) = 0$ can be represented as a smooth function of u and w, $x(u, w)$, both valid assumptions in this context.

The random vector w has a base distribution $\underline{\mathcal{P}}(w)$, i.e., a distribution exogenous to the plant. The objective function $\phi(u, w)$ has an associated posterior distribution $\overline{\mathcal{P}}(\phi(u, w))$, i.e., a distribution functionally dependent on u and x, from which its expected value will be maximized.

In a traditional stochastic framework, N_{samp} samples would be drawn from the distribution of w to convert the problem into a deterministic counterpart of the form

$$min_{u \in U} \tilde{V}(u) := \frac{1}{N_{samp}} \sum_{j=1}^{N_{samp}} [\phi(u, w_j)]. \tag{6.15}$$

However, this approach is computationally intensive because the number of samples must be overestimated in order to guarantee an appropriate solution accuracy. In this particular application, a limited number of samples can be estimated from the weather forecasting system. The BONUS reweighting scheme can instead be used to greatly reduce the number of samples required. This is done as follows:

Offline Computations

1. Independently distributed samples $j = 1, ..., N_{samp}$ are drawn for random variables $\hat{w}_j$ and inputs $\hat{u}_j$ covering the convex set U.
2. The design base density $\underline{P^d}(u, w)$ is estimated from these samples using kernel density estimation (KDE).
3. The objective function is evaluated for each sample $j = 1, ..., N_{samp}$ using a black-box model of the power plant and the objective function response, $\phi(\hat{u}_j, \hat{w}_j)$ is calculated.
4. The design current distribution $\overline{P^d}(u, w)$ is estimated using KDE.
5. Given the base distribution for the random variables $\underline{P}(w)$ and the design distributions $\underline{P^d}(u, w)$ and $\overline{P^d}(u, w)$, and objective value responses $\phi(\hat{u}_j, \hat{w}_j)$, $j = 1, ..., N_{samp}$, initialize decision variables u^k.
6. Optimality Check:
   a) Using the current iterate of decision variables u^k, a narrow normal distribution is defined around this point and samples u_j^k are drawn from it.
   b) Samples w_j^k are drawn from the available distribution.
   c) Samples are used to generate $\underline{P^k}(u^k, w)$ using KDE.
   d) The objective function is estimated using the reweighting formula

$$\tilde{V}(u^k) \approx \mathbb{E}_w[\phi(u * k, w)] \approx \sum_{j=1}^{N_{samp}} \omega_j^K (\phi(\hat{u}_j, \hat{w}_j), \tag{6.16}$$

   where the weights ω_j^k are obtained from

$$\omega_j^k = \frac{\frac{\underline{P^k}\left(u_j^k, w_j^k\right)}{\underline{P^d}(\hat{u}_j, \hat{w}_j)}}{\sum_{i=1}^{N_{samp}} \frac{\underline{P^k}\left(u_i^k, w_i^k\right)}{\underline{P^d}(\hat{u}_i, \hat{w}_i)}} \tag{6.17}$$

   and satisfy $\sum_{j=1}^{N_{samp}} \omega_j^k = 1$.
   e) The decision variables are perturbed $u^k \pm \delta u^k$ and reweighting is used to estimate the gradient $\nabla_u \tilde{V}(u^k)$ and the perturbed objective function $\tilde{V}(u^k + \delta u^k)$.

f) The gradient is checked against the Karush–Kuhn–Tucker (KKT) conditions

$$0 = \nabla_u V(u) + \overline{v} - \underline{v}, \tag{6.18}$$

$$0 \le \overline{v} \perp (\overline{u} - u) \ge 0, \quad 0 \qquad \le \underline{v} \perp (u - \underline{u}) \ge 0, \tag{6.19}$$

where $\underline{v}$ and $\overline{v}$ are the multipliers for the lower and upper bounds, respectively. If the KKT residual and complimentarity products are sufficiently small, the process is terminated. Otherwise, the process is continued in the next step.

7. SQP Step Computation:
   a) The gradient is used to compute the Hessian approximation H_k using the Broyden–Fletcher–Goldfarb–Shanno (BFGS) formula.
   b) The step δu is computed by solving the quadratic program (QP)

$$\min_{\Delta u} \delta_u \tilde{V}(u^k)^T \Delta u + \Delta u^T H^k \Delta u, \tag{6.20}$$

$$s.t. \quad u^k + \Delta u \in U. \tag{6.21}$$

   c) A new iterate is calculated as $u^{k+1} = u^k + \alpha \Delta u$ with $\alpha \in (0, 1]$, and the process is continued by returning to Step 6.

In this procedure, the computation of the objective function at each iteration requires only the values $\phi(\hat{u}_j, \hat{w}_j)$; the black box model is not re-run at each iteration of the optimization loop. This is a key benefit in this type of real-time application where the stochastic optimization problem must be solved many times.

6.5   Weather Forecasting and Uncertainty Quantification

Real-time forecasts and uncertainty information for the two weather factors of interest, ambient temperature and humdity, are computed using the numerical weather prediction (NWP) model WRF [53]. The WRF model is a state-of-the-art numerical weather prediction system designed to serve both operational forecasting and atmospheric research needs, and therefore has a comprehensive description of atmospheric physics that includes cloud parameterizations, land-surface models, atmosphere–ocean coupling, and broad radiation models. The complexity of the model, along with its extremely high resolution (up to 30 s of a degree), causes the WRF to be extremely computationally intensive. As such, special computational resources are required, and an implementation at Argonne National Laboratory is used for the work described in this chapter.

6.5.1   Ensemble Initialization

To initialize the NWP simulations, simulated atmospheric states are reconciled with experimental observations to create what are known as reanalyzed fields. The North

American Regional Reanalysis (NARR) data set is used to cover the North American continent with a resolution of 10 min of a degree and 29 pressure levels in 3 hour increments from 1979 until present.

Because of observation sparseness and the incompleteness of a numerical representation of atmospheric field dynamics, initial states are not exactly known and can only be represented statistically. Therefore, a distribution of the initial conditions, or "ensemble," is used to describe confidence in the knowledge of the initial state of the atmosphere. A normal distribution of the uncertainty field of the initial state is assumed, centered around the NARR field at the initial time, i.e., the expectation (first moment) exactly matches the NARR values.

The second statistical moment is approximated by the sample variance or pointwise uncertainty. The initial N_s-member ensemble field $\chi_s^{t_0} := \chi_s(t_0), s \in \{1...N_s\}$ is sampled from $\mathcal{N}(\chi_{NARR}, \sigma)$, a normal distribution centered at the NARR solution and with a standard deviation Σ:

$$\chi_s^{t_0} := \chi_{NARR} + \Sigma^{\frac{1}{2}} \zeta_x, \quad \zeta_s \sim \mathcal{N}(0, 1), \quad s \in \{1...N_s\}. \tag{6.22}$$

This is equivalent to perturbing the NARR field with $\mathcal{N}(0, \Sigma)$.

One challenge inherent in this methodology is the very large size of the correlation matrix. Therefore, it is not computed directly, and instead must be estimated by first building correlation matrices of forecast errors using the WRF model. These forecast errors are estimated using the NCEP method, which starts several simulations staggered in time in such a way that, for any point in time, two forecasts are available. The differences between the two staggered simulations is denoted as d_{ij}, where i is the point in space and j represents the pair of forecasts. The covariance matrix could then be approximated by $\Sigma \approx \mathbf{dd}^T$, but the size again makes this computationally intractable. Instead, correlations are assumed to be roughly similar across the continental US, greatly simplifying the calculation.

6.5.2   Ensemble Propagation

The initial state distribution is then propagated through the WRF. The resulting trajectories can then be assembled to obtain an approximation of the forecast covariance matrix:

$$\chi_x^{t_F} = \mathcal{M}_{t_0 \to t_F}(\chi_s^{t_0}) + \eta_s(t), \quad s \in \{1...N_s\}, \tag{6.23}$$

where $\chi_s^{t_0} \sim \mathcal{N}(\chi_{NARR}, \Sigma)$, $\eta_s \sim \mathcal{N}(0, \Sigma)$, and $\mathcal{M}_{t_0 \to t_F}(\cdot)$ represents the evolution of the initial condition through the WRF model from time t_0 to t_F. The initial condition is perturbed by the additive noise η that accounts for the various error sources during the model evolution.

In the application described in this chapter, it is assumed that the numerical model WRF is perfect, i.e., $\eta = 0$. Thus, given the exact initial conditions, the model is assumed to produce error-free forecasts. For long prediction windows, this is a strong assumption. However, the forecast windows in this application are restricted to no longer than one day in advance, thus making the assumption reasonable.

6.5.3   Validation of Weather Forecast

In Fig. 6.2, the average profile and 30 ensemble profiles are presented for the dry-bulb temperature and relative humidity for a day, respectively, in a random location in the Midwest region of the USA for June 1, 2006. The day-long profile is forecast at the location of a meteorological tower located near Chicago, IL ($41°42'04''$, $87°59'42''$). The crossed dots show the actual temperatures and humidity observed at the closest meteorological station as measured by the instruments mounted on the weather tower. The envelope surrounding the ensembles is the 95 % confidence region. Time zero represents 6 a.m. central time.

One can see that the projection for a day ahead is reasonably accurate from the WRF model, capturing well the trend and sensor readings. Temperature varies significantly throughout the day in the range of 288–299°K, while the relative humidity varies in the range of 20–90 %. Humidity is much higher at night, whereas temperature is much higher during the day, therefore the variables are autocorrelated. The temperature and RH uncertainties are the result of different NWP model forecasts for the same prediction window. While the temperature is relatively well modeled, the relative humidity (RH) is the result of the moisture transport in the atmosphere and complex physical interactions that lead to phase changes. Therefore, as expected, a relatively wider uncertainty estimate for RH and a fanning effect toward the end of the forecast window are observed. It should also be noted that, while uncertainty for temperature projections has been discussed in other studies, it was not possible to find projected uncertainty estimates for RH in the literature. These estimates are a contribution of this work.

The uncertainty envelopes for the ambient temperature are narrow. The uncertainty envelope for the relative humidity, on the other hand, is wide and can be as large as 30–40 % toward the end of the day. Both envelopes follow complex shapes due to the nonlinearity of the weather prediction model and current meteorological conditions. Toward the end of the day, the mean forecast does not predict well the shape of the humidity profile, but the uncertainty envelope covers this region reasonably well. This indicates the additional complexity in forecasting humidity. This also reflects the fact that the uncertainty envelope is unable to fully cover the sensor readings at all times.

The ambient temperature and relative humidity WRF ensembles are fitted to lognormal and uniform distributions, respectively. These are illustrated in Figs. 6.3 and 6.4. These were used to generate the design distribution $\underline{P}^d(u, w)$ and corresponding objective function posterior $\overline{P}^d(\varphi(u, w))$.

Fig. 6.2 Forecast (*thick line*) and ensemble profiles (*thin lines*) for dry-bulb temperature and relative humidity in Midwest US for June 1, 2006. Dots are real measurements from meteorological stations

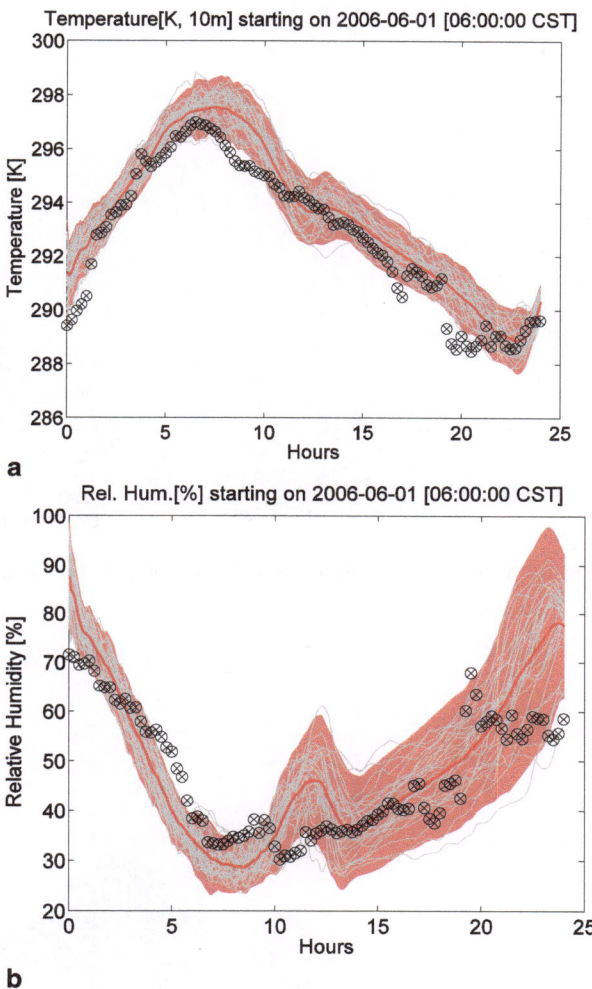

a

b

6.6 Application to Pulverized Coal Power Plant

Stochastic optimization and weather uncertainty quantification are combined in a numerical study of a PC power plant, the details of which are described in the previous chapter. Again, the power plant is modeled in Aspen Plus®, and is designed to generate 700 MW of electricity. A constraint on the maximum water intake of 9.1×10^5 kg/h is enforced. This represents the nominal consumption under average atmospheric conditions of 288 K and 60 % relative humidity.

In order to demonstrate the economic benefits of stochastic optimization over current practice, extensive simulations were performed using three settings.

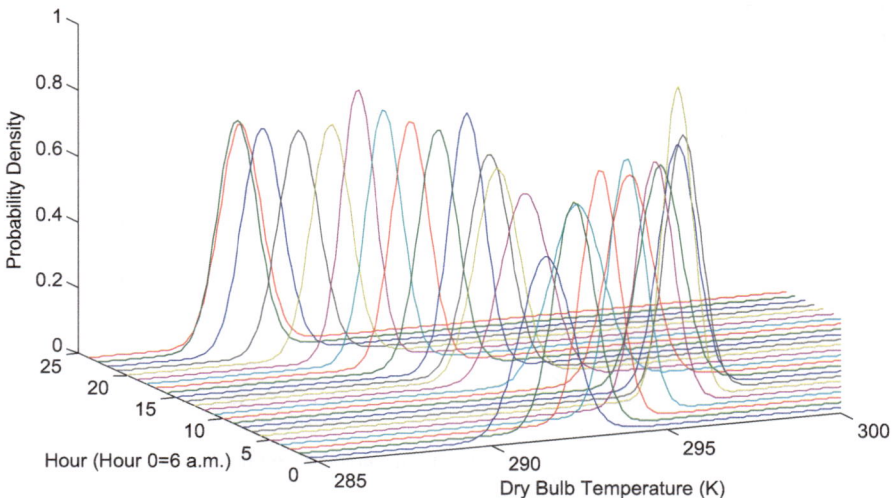

Fig. 6.3 Fit of ambient temperature WRF ensembles to lognormal distribution

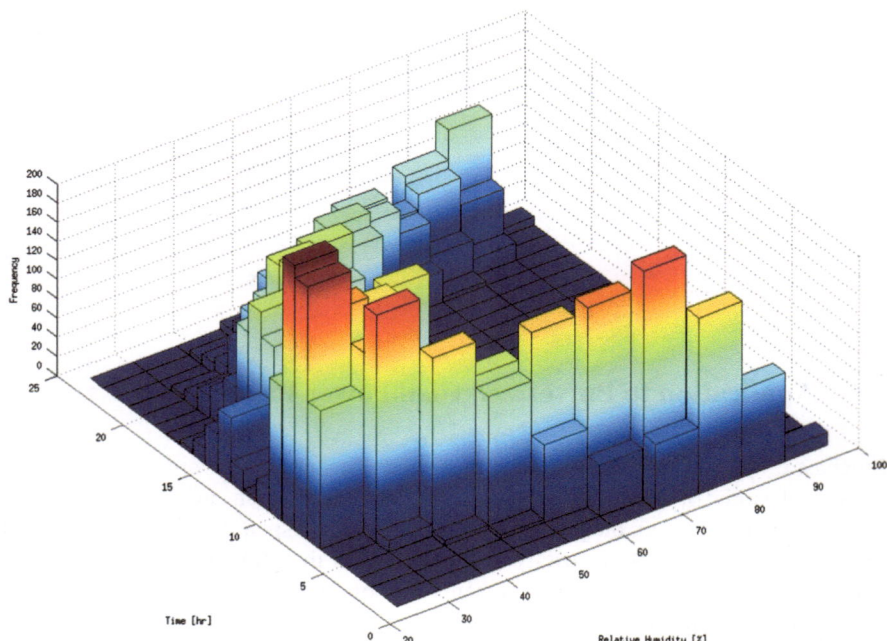

Fig. 6.4 Fit of relative humdity WRF ensembles to uniform distribution

1. **Deterministic profile (a):** the expected value of the maximum power was computed from the WRF ensemble information using day-average conditions for temperature and humidity.
2. **Deterministic profile (b):** the expected value of the maximum power was computed using the mean forecast values obtained from WRF.
3. **Stochastic profile:** the here-and-now stochastic optimization problem was solved using the WRF ensemble information to maximize the expected value of power during a time frame of 24 h with time steps of 1 h.

The maximum power profiles are shown in Fig. 6.5. Several findings are of interest:

- From deterministic profile (a), it can be seen that the power profile varies with weather conditions throughout the day from a low of 665 MW to a high of 715 MW. Consequently, the set points of the input parameters must be adjusted throughout the day to mitigate the effects of weather and water constraints.
- The highest power output is obtained at night when the ambient temperature is lowest, even if humidity is high. Low power is obtained around noon when the temperature is at its highest, regardless of humidity. This indicates that temperature is more important than humidity in determining cooling capacity. In addition, it is evident the power output is constrained by cooling capacity at high temperatures, but plant design capacity at low temperatures.
- From deterministic profiles (a) and (b), it can be seen that adjusting operating conditions according to the mean forecast can increase the maximum power output.
- Stochastic optimization gives a total output of 16,922 MWh, a performance gain of 245 MWh over deterministic profile (a) and 99 MWh over deterministic profile (b).
- Stochastic optimization does not fully mitigate the variability of power throughout the day. Rather, power varies from a low of 684 MW to a high of 729 MW, indicating the strong impact of weather and corroborating the fact that the plant is constrained by cooling capacity.
- At hours 15 and 23, the stochastic profile is lower than that of deterministic profile (b). This is a result of the estimation error introduced from the small number of samples. However, the stochastic solution is consistently better than the deterministic one, so it may be concluded that the number of samples used was sufficient to obtain useful solutions. This, in turn, implies that weather uncertainty information is accurate and the BONUS reweighting scheme is efficient in identifying quality solutions.

Overall, these results indicate that, for a given water intake constraint, power can be maximized by using stochastic optimization. Further, as water constraints are tightened, it is possible to maintain high output levels with stochastic optimization, thus the methodology can be used in reducing water consumption, though tightening the water intake constraint will make the power profile more sensitive to weather conditions.

The 245 MWh gain, projected over 100 days in a year (the approximate number of days subject to summer conditions) results in an increase of 24.5 GWh. At a

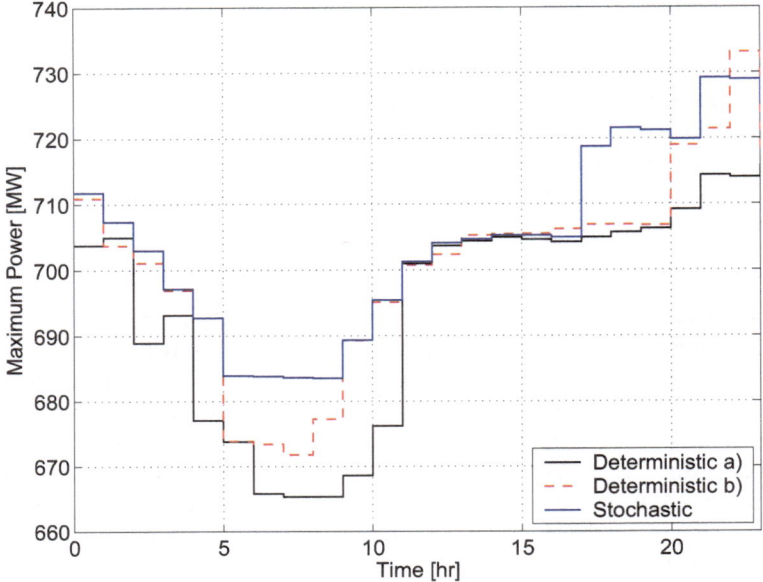

Fig. 6.5 Maximum power profiles for deterministic and stochastic approaches

price of $ 100/MWh, this translates into $ 2,450,000. While this is a rough estimate, void of seasonal effects and differing constraint scenarios, the order of magnitude is informative.

A traditional stochastic optimization approach with no reweighting would have used 355,200 runs of the ASPEN Plus® model to solve the problems over the 24 h time frame. At approximately 10 min per run, this would have required 2400 days. By contrast, using the BONUS reweighting scheme, only 600 runs of the ASPEN Plus® model were required, a savings of 99.8 %. Therefore, the BONUS reweighting scheme makes stochastic optimization possible for this problem that would otherwise be tractably infeasible.

6.7   Summary

A real-time stochastic optimization framework for maximizing power output in a PC power plant is shown to provide significant gains in power output, as compared to current state optimization. This approach integrated detailed physical weather and power plant simulation models to assess the effects of ambient weather conditions on plant performance. While infeasible in a traditional stochastic framework, the BONUS reweighting scheme makes the problem tractable and allows for gains of approximately $ 2.45 million per year over current practice.

Notations

$\mathbb{E}_w[\cdot]$	expected output power
$J(x, u, w)$	denotes the plant power output
N_{samp}	number of samples taken of the uncertain parameters
$P(t)$	total amount of power produced by the plant at time t
P^{max}	maximum power output under design specifications (nominal capacity)
$Q_c(u_c, w)$	cooling system capacity
Q_c^{max}	maximum cooling capacity under ideal weather conditions
$Q_p(t)$	thermal load at time t
Q_p^{max}	maximum cooling demand
x_c	set of cooling system state variables
$x_p(t)$	set of system state variables
u_c	set of cooling system input parameters
$u_p(t)$	set of input variables
U	set $U := \{u | \underline{u} \leq u \leq \overline{u}\}$,
δu	step size in the SQP algorithm
	where $\underline{u}$ and $\overline{u}$ are the upper and lower bounds, respectively
	$V(u_{HN}) := \mathbb{E}_w[J(x(w), u_{HN}, w)]$
	optimal value of expected power
w	set of weather conditions

Greek letters

σ	standard deviation
$\phi(u, w)$	value of $J(x(w), u, w)$ at a given
	$x(w), w, u$ that solves the model $f(x(w), u, w) = 0$
Ω	set of values over which w has support

Chapter 7
Sensor Placement Under Uncertainty for Power Plants

7.1 Introduction

This chapter demonstrates the use of the BONUS method, in combination with kernel density estimation (KDE), to calculate Fisher information. This concept is then applied to the problem of sensor placement in an integrated gasification combined cycle (IGCC) power plant, and how BONUS significantly reduces computational resources while contributing to an appropriate solution is shown. This chapter is derived from the work by Lee and Diwekar [28].

7.1.1 The Integrated Gasification Combined Cycle Power Plant

The IGCC power plant is a cleaner way of getting electricity from coal compared to the pulverized coal (PC) plant described in earlier chapters. IGCC consists of three main elements: the air separation unit (ASU), the gasification plant, and the power block, as shown in Fig. 7.1 [35]. Power is produced in the IGCC power plant as follows:

1. The ASU separates ambient air into oxygen (O_2) and nitrogen (N_2). The oxygen is used primarily to produce fuel gas in the gasification plant, while most of the nitrogen is used to dilute fuel gas and reduce nitrous oxide (NO_x) levels in the power plant's combustion turbine.
2. The gasification plant converts coal or other solid fuel (e.g., petroleum coke or biomass) into fuel gas and high pressure steam by reacting with the O_2 produced by the ASU in several steps.
   a) The coal is received and stored in the plant in the form of coal fines, finely powdered solid material.
   b) Coal fines are mixed with water and ground into a viscous slurry.
   c) The coal slurry and oxygen react in the gasifier to produce:

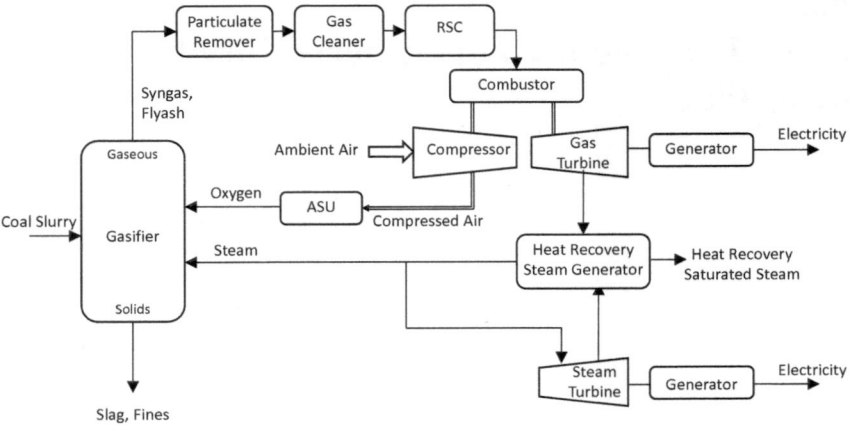

Fig. 7.1 The integrated gasification combined cycle power plant

- Syngas, a synthetic gas composed of hydrogen (H_2), water vapor (H_2O), carbon monoxide (CO), and carbon dioxide (CO_2)
- Slag, the residual mineral matter from coal not converted to syngas, which afterward flows down the gasifier walls, solidifies into an inert glassy frit with little carbon content, and is removed as waste
- Flyash, partially gasified residual carbon that exits the gasifier within the syngas stream

    Both slag and flyash are undesirable byproducts of the reaction. The gasifier typically operates at a temperature and pressure around 1645 K and 2760 KPa.

    d) The syngas is cooled in a radiant syngas cooler (RSC), then passed through a high pressure steam generator and gas cooler. The efficiency of this steam generation step may be improved by employing hot gas desulfurization to reduce nitrous oxide (NO_x) emissions.

    e) Intensive water scrubbing removes flyash and other particulate matter from the syngas.

    f) COS is converted to H_2S and removed from the syngas.

    g) Selective catalytic reduction (SCR) removes NO_x from the process.

3. The power block generates electricity from the fuel gas, nitrogen, and high pressure steam.

    a) The fuel gas powers a combustion turbine.

    b) A heat recovery steam generator (HRSG) uses the gas turbine exhaust gas to generate both high and low pressure steam.

    c) The high and low pressure steam powers additional turbines, generating electricity.

A high efficiency combined cycle helps lower SO_2, NO_x, and particulate levels, reducing the environmental impact of the IGCC plant power generation process.

7.1.2   Measurement Uncertainty

Monitoring every process variable contained in the IGCC plant operations using a complete network of sensors would prove to be both costly and, due to the inability of obtaining measurements within harsh environments, technically infeasible. At the same time, however, controlling the operating conditions is essential to maintaining the efficiency of the power generation process. Therefore, some process variables must be estimated from related process variables for which direct measurements are more easily taken.

Gasifier temperature provides an illustrative example. Because the gasifier operates at extreme temperature and pressure, standard thermocouples cannot be used to take direct measurements. This makes it difficult to both determine and maintain a target operating temperature. However, the durability of the gasifier decreases at higher temperatures, while slag output increases at lower temperatures, thus a variation in either direction from the optimal temperature increases both cost and environmental impact. Therefore, gasifier temperature must be inferred by measuring related process variables. In this case, the methane production rate depends on both gasifier temperature and fuel composition, allowing measurements of the methane production rate and fuel composition, which are more easily obtained in practice, to be used to estimate gasifier temperature.

The large number of process variables and the complex relationships among them generate a significant challenge in determining which variables should be directly measured and which should be estimated, or "indirectly measured." Each direct measurement requires the use of a sensor, and the network of online sensors is defined as the full set of sensors used. The problem herein is to design a network of online sensors so as to minimize the overall costs, including purchase, deployment, and maintenance, associated with that network, while enabling a sufficient level of process control. As part of a stochastic optimization problem, the decision to either observe or estimate each process variable results from the uncertainty surrounding the true values of the process variables in the form of system and measurement noise.

To solve the cost-minimization problem, the IGCC power plant is modeled in Aspen Plus® to quantify the variability of downstream process variables as a result of variability in a set of input process variables, such as coal and oxygen flow rates, gasifier temperature, and gasifier pressure. Using the known measurement distributions of online sensors that are a priori assumed to be part of the sensor network, the downstream process variability is captured using Fisher information, as detailed in the next section. The Fisher information is then used within the objective function to determine which of the downstream process variables should be observed or physically measured through the placement of candidate sensors.

7.2   Fisher Information and Its Use in the Sensor-Placement Problem

Fisher information is a statistical measure established in the field of information theory by Ronald Fisher [11]. For a set of independent and identically distributed (IID) observations, $x_1, x_2, ..., x_n$, resulting from n outcomes of a random variable, $X = X_i, i = 1, 2, ..., n$, Fisher information captures the amount of information the set of observations contains about some unknown parameter, θ_x, upon which the probability distribution of X, $p_x(x)$, depends. It does this by quantifying the expected change in the distribution due to a change in the parameter value, θ_x. The expression for Fisher information, I_x, is commonly given as [12]

$$I_x(\theta_x) = E^X \left[\left(\frac{1}{p_{X|\Theta_x}(X|\theta_x)} \frac{\partial p_{X|\Theta_x}(X|\theta_x)}{\partial \theta_x} \right)^2 \right], \qquad (7.1)$$

where the distribution $p(X|\theta_x)$ is the likelihood of x given the parameter θ_x.

In the sensor-placement problem, a high level of Fisher information for a downstream variable indicates the ability to accurately estimate the value of an upstream variable on which the downstream variable depends. Because Fisher information is additive $(I_{X,Y}(\theta) = I_X(\theta) + I_Y(\theta))$, a single Fisher quantity may be calculated for the entire system. Thus, the goal is to decrease the overall sensor cost by determining the optimal sensor locations to maximize the amount of information about the system's true state.

7.3   Computation of Fisher Information

Using the Aspen Plus® environment, a comprehensive model of the highly nonlinear IGCC process is used to simulate the steady-state performance of the ASU, gasifier, and power generation processes. This Aspen model is used to estimate the set of unmeasured variables using the data acquired from the process variables directly measured through the network of sensors physically deployed within the plant.

Let S^{in} be the set of input variables, including coal and oxygen flow rate. Each variable in S^{in} follows a uniform distribution centered at its nominal value. A set of N_{samp} input variable operating conditions is generated using Hammersley sequence sampling, and the IGCC process is simulated in Aspen N_{samp} times. Each simulation generates a corresponding vector of points, S^{out}, that includes both intermediate and output process variables, such as syngas temperature and mass flow rate. S^{out} captures the nonlinear effects of the IGCC process, and the full set of S^{out} vectors generated from repeated simulations captures the variability of downstream process variables resulting from a uniformly distributed set of input variable sample points. Thus, a probability distribution can be generated for each intermediate and output variable that captures the variation in that variable due to variations in the input variables and the nonlinearity of the process behavior.

7.3.1 Reweighting Using the BONUS Method

Each time the network of sensors is altered, i.e., a sensor is added or removed from the online network, the underlying distributions of the process variables are altered, requiring a new computation of the Fisher information about each process variable. In this section, the BONUS algorithm is used to compare samples of the input variables taken from a uniformly distributed sample space to those taken from a new reference distribution in order to create a set of distribution weights that can be used to reweight the distribution functions of the intermediate and output variables. This reweighting approach eliminates the need to resimulate the IGCC process behavior in Aspen Plus for every possible combination of online sensors, thereby significantly reducing the overall computational time.

The BONUS reweighting scheme is implemented as follows:

1. Let $f_0(x_i)$ be the probability density function (PDF) associated with the base input distribution for the input variable $x_i, i = 1, 2, ..., S^{in}$.
2. A set of N_{samp} sample points uniformly distributed across a d-dimensional samples space is used to perform N_{samp} simulations of the IGCC process to generate $F_0(y_j)$, the base cumulative distribution function (CDF) associated with the intermediate or output variable $y_j, j = 1, 2, ..., S^{out}$, where $y_j = h(x_1, x_2, ..., x_{S^{in}})$ is the nonlinear transformation from the set of input variables, S^{in}, to the downstream variable y_j at iteration 0.
3. A new input distribution is defined, representing a change in sensor placement, such as a sensor placed at the location of this input variable. The redefined distribution, $f_t(x_i)$, at iteration t is used to create a set of weights

$$W_t(x_i) = \frac{f_t(x_i)}{f_0(x_i)}, i = 1, 2, ..., S^{in} \tag{7.2}$$

that gives the likelihood ratio between the redefined and base distributions.
4. Given that the input variables act independently, the weights are used to construct the resulting distributions for the downstream intermediate and output variables at iteration t by multiplying the associated weights, $W_t(x_i)$, with the base distribution $f_0(y_j)$:

$$f_t(y_j) = f_0(y_j)\Pi_{i=1}^{S^{in}}(1 + \gamma_{ij}(W_t(x_i) - 1)), j = 1, 2, ..., S^{out}, \tag{7.3}$$

where $\gamma_{ij} = 1$ if variable y_j is downstream of x_i and $\gamma_{ij} = 0$ if it is not.
5. The distribution is then normalized using

$$\hat{f}_t(y_j) = \frac{f_t(y_j)}{\sum_{n=1}^{N_{samp}} f_t(y_j(n))\frac{y_j(n+1)-y_j(n-1)}{2}}. \tag{7.4}$$

This reweighting approach can also be used when a sensor is placed at the location of an intermediate process variable to construct the resulting change in distributions of corresponding downstream variables. By eliminating the need to generate a new set of N_{samp} sample points through simulation of the IGCC process at each iteration, t, the BONUS reweighting algorithm provides an efficient method for calculating the Fisher information resulting from many different configurations of the online sensor network. Further, various underlying distributions corresponding to sensor accuracies can be readily analyzed without increasing the computational burden, and this approach can also be used for unmeasured disturbances to an input variable, such as a change in coal quality.

7.3.2 Calculating the Fisher Information from Kernel Density Estimation

As discussed in Chap. 3, KDE is a nonparametric method of estimating the PDF of a random variable based on a finite data sample. In this case, the finite data sample consists of the set of operating parameters estimated in Aspen PLUS for each input sample. The KDE technique estimates the PDF through the use of following formula:

$$p(y_n) = \sum_{m=1}^{N_{samp}} \frac{1}{\sqrt{2\pi}} exp \left(- \left(\frac{y_n - y_m}{h} \right)^2 \right), \quad (7.5)$$

at each sample point $y_n, n = 1, 2, ..., N_{samp}$, where σ^2 is defined as the variance of the set of samples $\{y_1, y_2, ..., y_N\}$ and $h = 1.06\sigma/N_{samp}^{1/5}$.

Assume that the shift-invariant property holds for a small $\epsilon > 0$ change in the parameter θ_y (the mean value of a given y), i.e., $p(y_n \pm \epsilon)$ can be calculated from (7.5) by replacing y_n with $y_n \pm \epsilon$ on the right side of the equation. This is a viable assumption at IGCC process operating conditions near their means, as the plant is operated within a chemically stable region. Once the kernel density functions $p(y_n + \epsilon)$ and $p(y_n - \epsilon)$ are calculated from (7.5), they can be used to generate an approximation of the first-order derivative, $\partial p(y_n)/\partial \theta_y$, given by

$$\frac{\partial p(y_n)}{\partial \theta_y} \approx \frac{(p(y_n + \epsilon) - p(y_n - \epsilon))}{2\epsilon}. \quad (7.6)$$

The Fisher information is then obtained by substituting (7.6) into the discrete approximation of (7.1) to obtain

$$I_y(\theta_y) = \sum_{n=1}^{N_{samp}} (y_n - y_{n-1}) \frac{\left(\partial p(y_n)/\partial \theta_y \right)^2}{p(y_n)}, \quad (7.7)$$

which constructs a series of right-hand rectangles at $\frac{\left(\partial p(y_n)/\partial \theta_y \right)^2}{p(y_n)}$ to approximate the integral function in the expectation.

The following section applies Fisher information as a metric of observation order (the degree to which a given sensor network can monitor and control the system) within an optimization problem for placing sensors in various locations throughout the IGCC plant, subject to sensor cost constraints.

7.4   The Optimization Problem

The objective of the sensor-placement problem is to maximize the amount of information about the IGCC process from a network of sensors, given a set of budget constraints. Because it is desirable to minimize the variability of the unmeasured process variable estimations, the Fisher information should be maximized.

The resulting optimization problem is a nonlinear stochastic (binary) integer problem where the objective function consists of the overall Fisher information (with the goal of maximization) and the constraints consist of limits on the cost of sensor placement. Formally, this is given as

$$\max_{w_j \in \mathbb{W}} \sum_{j=1}^{S^{out}} f_j(\mathbf{w}, \mathbf{Y}) w_j, \tag{7.8}$$

$$s.t \sum_{j=1}^{S^{out}} C_j w_j \le B, \tag{7.9}$$

$$w_j \in 0, 1, \quad j = 1, 2, ..., S^{out}, \tag{7.10}$$

where C_j is the cost associated with the purchase, deployment, and maintenance of sensor j and B is the total sensor budget. The binary variable w_j represents the decision to place or not place sensor j in the network of online sensors, with 0 representing the absence of sensor j and 1 representing its presence, and $\mathbb{W}$ constitutes the set of all feasible sensor networks that is given.

7.4.1   Defining the Objective Function

The objective term $f_j(w, Y)$ is a function of the Fisher information resulting from the network of sensors, $\mathbf{w} = \{w_j \in \{0, 1\}, j = 1, 2, ..., S^{out}\}$ and the set of random variables $\mathbf{Y} = \{Y_j, j = 1, 2, ..., S^{out}\}$ associated with the measurement uncertainties in the intermediate and output process variables. This function is designed by first assuming that the information related to a process variable is always greater if a sensor is placed online at that specific location (i.e., more is known about Y_j when $w_j = 1$). Let $I_{Y_j}^s(\theta_{y_j} | w_k = 1)$ represent the Fisher information of θ_{y_j} resulting from a sensor placed at location $K = 1, 2, ..., S^{out}$, and let $I_{Y_j}^{ns}(\theta_{y_j} | w_k = 0), k = 1, 2, ..., S^{out}$ represent the Fisher information of θ_{y_j} resulting from no sensors placed in the network of intermediate and output variables, such that $I_{Y_j}^s(\theta_{y_j} | w_k = 1) \ge I_{Y_j}^{ns}(\theta_{y_j}), j = 1, 2, ..., S^{out}$ (this inequality states that the information about variable j that is known when there is a sensor measuring variable k is greater than or equal to the information about variable j that is known when there is *not* a sensor measuring variable k). A candidate objective function can then be defined as

$$f_j^A(\mathbf{w}, \mathbf{Y}) = 1 - \frac{I_{Y_j}^{ns}(\theta_{y_j})}{I_{Y_j}^s(\theta_{y_j}|w_j = 1)}, \tag{7.11}$$

where $0 \leq f_j^A(\mathbf{w}, \mathbf{Y}) \leq 1$. This function normalizes the Fisher information for each process variable between zero and one. Values of $f_j^A(\mathbf{w}, \mathbf{Y})$ close to zero correspond to the smallest change in information gained from placing a sensor at location j, while values close to one correspond to the largest change. It is therefore possible to optimize the placement of sensors across many variable attributes, including mass-flow, temperature, and pressure, for example, by determining which set of sensors provides the largest total gain in estimation of the dynamic system.

However, this function does not capture the potential effects of placing a sensor in the network upstream of location j. If location k is upstream of location j (i.e., Y_j is dependent on Y_k), then information gained by placing a sensor at location k increases the amount of information available about Y_j. A second candidate objective function that takes this into account is

$$f_j^B(\mathbf{w}, \mathbf{Y}) = \sum_{k=1}^{S^{out}} \left(1 - \frac{I_{Y_j}^{ns}(\theta_{y_j})}{I_{Y_j}^s(\theta_{y_j}|w_k = 1)}\right), \tag{7.12}$$

which captures the overall effect that placing (or not placing) a sensor has on all other process variables by summing the resulting information gained at all locations by placing a sensor at location k. The Fisher information is given as $I_{Y_j}^s(\theta_{y_j}|w_k = 1) = I_{Y_j}^{ns}(\theta_{y_j})$ if variable j is not downstream of variable k, and it can be seen that, in this case, the right-hand side of Eq. (7.12) reduces to zero. Otherwise, if j *is* downstream of variable k, $I_{Y_j}^s(\theta_{y_j}|w_k = 1)$ can be computed using the BONUS reweighting scheme.

7.4.2   The IGCC Power Plant

For the IGCC power plant studied, a set of eight sensors, S^{in}, measures the input process variables given in Table 7.1. The objective is to determine the placement of sensors across a set of 24 sensors, S^{out}, measuring intermediate and output variables $y_1, y_2, ..., y_{24}$ given in Eq. 7.2, as well as the nominal operating conditions. A schematic of potential sensor locations is given in Fig. 7.2.

For each intermediate and output process variable, three types of sensors are assumed to be available, with accuracies (six standard deviations) of $\pm 5\%$, $\pm 2.5\%$, and $\pm 1\%$, with sensors of higher accuracy incurring a higher cost than those of lower accuracy. The optimization problem is therefore slightly modified to include the consideration of multiple sensor types. Let the binary variable $w_{j,\tau} = 1(0)$ correspond to the decision to place a sensor of type $\tau = 1, 2, 3$ at location j. The problem can then be formulated as

Table 7.1 Input process variables

x_i	Description	Nominal	Units
1	Oxygen flow rate entering ASU	157,392	kg/h
2	Coal slurry flow rate	192,922	kg/h
3	Air flow rate to gas turbine compressor	2,962,683	kg/h
4	Recycled HRSG steam temperature	414	K
5	Recycled HRSG steam pressure	526	KPa
6	Recycled HRSG water temperature	369	K
7	Gasifier temperature	1644	K
8	Gasifier pressure	2806	KPa

ASU air separation unit, *HRSG* heat recovery steam generator

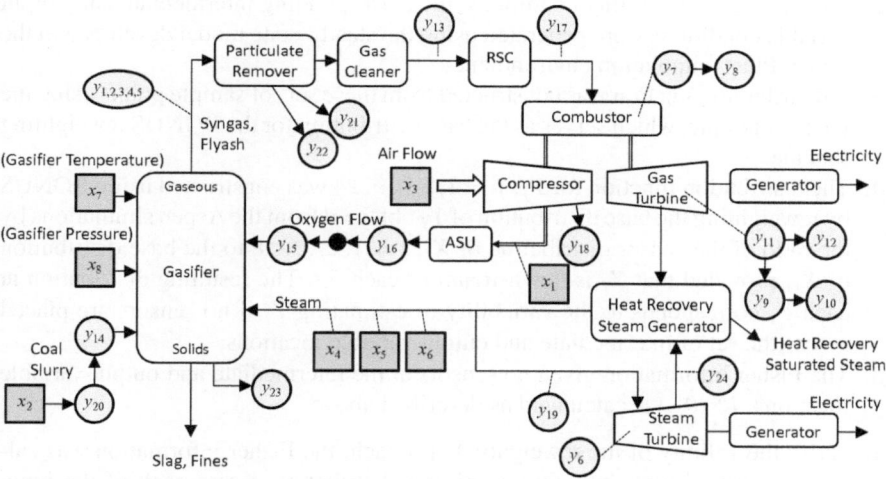

Fig. 7.2 Potential sensor locations in the IGCC power plant. *IGCC* integrated gasification combined cycle

$$\max_{w_{j,\tau} \in \mathbb{W}} \sum_{\tau=1}^{3} \sum_{j=1}^{24} f_{j,\tau}(\mathbf{w}, \mathbf{Y}) w_{j,\tau} \tag{7.13}$$

$$s.t \sum_{\tau=1}^{3} \sum_{j=1}^{24} C_{j,\tau} w_{j,\tau} \leq B, \tag{7.14}$$

$$\sum_{\tau=1}^{3} w_{j,\tau} \leq 1, \quad j = 1, 2, ..., 24 \tag{7.15}$$

$$w_{j,\tau} \in \{0, 1\}, \quad j = 1, 2, ..., 24, \ \tau = 1, 2, 3, \tag{7.16}$$

where $f_{j,\tau}(\mathbf{w}, \mathbf{Y})w_{j,\tau}$ is a function of the Fisher information when a sensor of type τ is placed at location j. Constraint (7.15) ensures that no more than one type of sensor is used at each location.

7.4.3 Problem Approach

The problem was approached in five steps:

1. A set of $N_s = 800$ operating conditions was generated across a uniform 8-dimensional sample space, corresponding to a set of 8 input variables varied $\pm 10\%$ of their nominal operating conditions using the Hammersley sequence sampling method.
2. For each set of operating conditions, the corresponding intermediate and output variable conditions were generated using the steady-state model developed in the Aspen Plus® simulation environment.
3. A distribution function was constructed from these sets of sample points using the KDE technique, which serves as the base distribution for the BONUS reweighting scheme.
4. The distribution function for $Y_j, j = 1, 2, \ldots, 24$ was constructed using BONUS by reweighting the base distribution of Y_j obtained from the Aspen simulations by the ratio of the sensor distribution of $X_i, i = 1, 2, \ldots, 8$ to the base distribution of X_i, provided that Y_j is downstream of each X_i. The resulting distribution at each Y_j corresponds to the variability of estimating Y_j if no sensors are placed across the set of intermediate and output variable locations.
5. The Fisher information given no sensors at the intermediate and output variable locations, $I_{Y_j}^{ns}(\theta_{y_j})$ is calculated as described above.

To verify the validity of the reweighting approach, the Fisher information was calculated two ways: first, by using a uniform distribution across each of the input variables as the input to the Aspen Plus® simulation, followed by use of the BONUS reweighting scheme, and second, by using a normal distribution across each of the input variables as the input to the Aspen Plus® simulation. There was no significant difference in the Fisher information calculated under each of the two methods. This is because the number of sample points and the sampling scheme used ensured adequate coverage of the 8-dimensional space, and the reweighting approach undergoes only one iteration when computing the Fisher information for a given set of input variable distributions. Thus, it is evident that the BONUS reweighting scheme is a useful approach for comparing sensor networks with contrasting variability, rather than rerunning the resource-intensive simulation in Aspen Plus® (Table 7.2).

Table 7.2 Intermediate and output process variables

y_j	Description	Stream[a]	Nominal	Units
1	Gasifier syngas flow rate	RXROUT	393,475	kg/h
2	Syngas CO flow rate	RXROUT	224,637	kg/h
3	Syngas CO_2 flow rate	RXROUT	88,051	kg/h
4	Syngas temperature	RXROUT	1644	K
5	Syngas pressure	RXROUT	2806	KPa
6	Low pressure steam turbine temperature	TORECIR	369	K
7	Gas turbine combustor burn temperature	POC2	1628	K
8	Gas turbine combustor exit temperature	POC3	1533	K
9	Gas turbine high pressure exhaust stream temperature	GTPC3	621	K
10	Gas turbine low pressure exhaust stream temperature	GTPC9	404	K
11	Gas turbine expander output temperature	GTPOC	872	K
12	Fluegas flow rate exiting gas turbine expander	6X	5,760,623	kg/h
13	Syngas flow rate after solids removal	RAWGAS	467,200	kg/h
14	Coal slurry flow rate entering gasifier	COALD	21,170	kg/h
15	Oxygen flow rate into gasifier	O2GAS	157,452	kg/h
16	Oxygen flow rate exiting ASU	GASIFOXY	157,452	kg/h
17	Acid gas flow rate	FUEL1	344,996	kg/h
18	Gas turbine compressor leakage flow rate	XCLEAK	2052	kg/h
19	Flow rate into high pressure steam turbine	TOHPTUR	621,421	kg/h
20	Coal slurry feed flow rate	COALFEED	192,922	kg/h
21	Slag extracted from syngas	SLAG	15,805	kg/h
22	Fines extracted from syngas	FINES	5363	kg/h
23	Gasifier heat output	QGASIF	2.47e7	Btu/h
24	Recycled HRSG[b] steam heat output	QRDEA	3.27e8	Btu/h

[a] Stream notation refers to DOE/NETL model [35]
[b] *HRSG* heat recovery steam generator

7.4.4   Results

Table 7.3 lists the computed objective values using the normalized function $f_j^B(\mathbf{w}, \mathbf{Y})$ from Eq. 7.12. As the sensor accuracy at a location increases, the value of f_j^B at that location increases due to the decrease in measurement variability, resulting in an increase in information pertaining to the true value of the variable at that location. Note that some variables, such as gasifier syngas flow rate (y_1) and fluegas flow rate exiting gas turbine expander (y_{12}), exhibit large increases in information when a

Table 7.3 Computed
objective values, f_j^B, for each
sensor type

Sensor j	Low accuracy	Medium accuracy	High accuracy
1	0.9100	8.6612	10.6078
2	9.8488	10.7561	10.9649
3	10.5601	10.8862	10.9898
4	7.8290	10.1407	10.8627
5	7.8989	10.1472	10.8613
6	0.1036	0.7760	0.9643
7	4.6106	5.6794	5.9470
8	3.7799	4.7002	4.9529
9	1.9262	1.9832	1.9981
10	0.9940	0.9989	1.0002
11	2.5901	2.9110	2.9845
12	0.0002	0.7054	0.9531
13	0.9188	6.2690	7.6865
14	12.4675	15.8420	16.8025
15	12.4553	15.8393	16.8083
16	13.3944	16.8241	17.8059
17	3.6553	6.2014	6.8691
18	0.9389	0.9849	0.9978
19	0.0002	0.0002	0.0002
20	13.4061	16.8267	17.8000
21	0.7492	0.9375	0.9902
22	0.7492	0.9375	0.9902
23	1.0002	1.0002	1.0002
24	0.0002	0.0002	0.0002

more accurate sensor is used, while others, such as gas turbine low pressure exhaust
steam temperature (y_{10}) and flow rate into high pressure steam turbine (y_{19}), show
little improvement in Fisher information from use of a more accurate (and therefore
costly) sensor.

Consider the case in which the total budget is $B = \$1,500,000$. The solution to
the optimization problem places a network of low accuracy sensors at locations y_2,
y_3, y_5, y_9, and y_{11}, and medium accuracy sensors at y_1, y_{14}, y_{15}, y_{16}, y_{17}, and y_{20}
(thus y_4, y_6, y_7, y_8, y_{10}, y_{12}, y_{13}, y_{18}, and y_{19} are not directly measured). The resulting
standard deviation in the IGCC power plant production and gasifier performance is
provided in Table 7.4, in comparison with the standard deviation resulting from the
baseline case in which no sensors are deployed across the intermediate and output
process variable location. The significant reduction in variability for both gas turbine
power production and total plant power production is immediately obvious.

Table 7.4 Measurement variation of the integrated gasification combined cycle (*IGCC*) power production and gasifier performance using the optimal sensor network versus no sensors deployed

IGCC power production	Nominal	Standard deviation Optimal (no sensors)	Units
Gas turbine power production	424.94	2.26 (43.11)	MWE
Steam turbine power production	251.97	0.71 (0.71)	MWE
Miscellaneous power consumption	− 67.41	0.25 (4.62)	MWE
Auxiliary power production	18.29	1.35 (1.35)	MWE
Total plant power production	591.22	2.16 (43.73)	MWE
Gasifier performance	Nominal	Standard deviation Optimal (no sensors)	Units
Oxygen flow rate	157,452	655 (13,386)	kg/h
Coal flow rate	192,922	803 (10,874)	kg/h
Slag flow rate	15,805	46 (1097)	kg/h
Fines flow rate	5363	16 (372)	kg/h
Syngas temperature	1645	370 (370)	K
Syngas pressure	2806	23 (234)	KPa

7.5   Summary

The use of the BONUS reweighting scheme can significantly reduce the computational resources required to calculate Fisher information, here used as a measurement of the variability of system parameters, given limitations on direct measurement of variables. This greatly improves the tractability of a nonlinear, stochastic integer program used to design a network of online sensors in an IGCC power plant, seeking to minimize variability while respecting budgetary constraints. In the case presented, measurement variability of total plant power production was reduced by over 95 %.

Notations

B total sensor budget
C_j cost associated with the purchase, deployment, and maintenance of sensor j
$f_0(x_i)$ probability density function (PDF) associated with
             the base input distribution for the input variable $x_i, i = 1, 2, \ldots, S^{in}$
$F_0(y_j)$ base cumulative distribution function (CDF)
             associated with the intermediate or output variable $y_j, j = 1, 2, \ldots, S^{out}$
$f_t(x_i)$ redefined input distribution
h band width
I_x Fisher information
N_{samp} number of input scenarios generated using Hammersley sequence sampling

S^{in} set of input variables, including coal and oxygen flow rate
S^{out} set of intermediate and output process variables,
                such as syngas temperature and mass flow rate
t iteration
w_j decision variable for placement of sensor j in the network of online sensors,
                with 0 representing the absence of sensor j and 1 representing its presence
$\mathbb{W}$ set of all feasible sensor networks
$W_t(x_i)$ weight used in the BONUS algorithm that
                gives the likelihood ratio between the redefined and base distributions
x_i observations
X random variable

Greek letters
γ_{ij} position indicator equal to 1
                if variable y_j is downstream of x_i and $\gamma_{ij} = 0$ if it is not

Chapter 8
The L-Shaped BONUS Algorithm

This chapter is based on a paper by Shastri and Diwekar [49] . A variant of BONUS is presented here to solve multistage stochastic programming problems with recourse. In stochastic programming problems with recourse, there is action (x), followed by observation, and then recourse r. In these problems, the objective function has the action term, and the recourse function is dependent on the uncertainties and recourse decisions. The recourse function can be a discontinuous nonlinear function in x and r space. A general approach behind the L-shaped method is to use a decomposition strategy where the master problem decides x and the subproblems are solved for the recourse function (Fig. 8.1). The method is essentially a Dantzig–Wolfe decomposition [6] (inner linearization) of the dual or Bender's decomposition of the primal. This method is due to Van Slyke and Wets [58], for stochastic programming also considers feasibility questions of particular relevance in these recourse problems. Consider the generalized representation of the recourse problem shown below, where the first term depends only on x, and R is the recourse function which depends on decision variables, x, recourse variables, r, and uncertain variables, u.

$$\text{Minimize } Z = f(x) + R(x, r, u) \qquad (8.1)$$
$$x$$

subject to

$$h(x, r) = 0 \qquad (8.2)$$
$$g(x, u, r) \leq 0 \qquad (8.3)$$

Figure 8.1 shows the decomposition scheme used in the L-shaped method. In the figure, the master problem is the linearized representation of the nonlinear objective function (containing the recourse function) and constraints. The master problem provides the values of the action variables x (x^*) and obtains the lower bound of the objective function. In general, the multistage recourse problems involve equality constraints relating the action variables x to the recourse variables r as in the generalized representation. These constraints are included as inequalities (feasibility cuts) in terms of the dual representation (including Lagrange multipliers λ) obtained by solving the following feasibility problem for each constraint. The feasibility cut

© Urmila Diwekar, Amy David 2015   95
U. Diwekar, A. David, *BONUS Algorithm for Large Scale Stochastic Nonlinear Programming Problems,* SpringerBriefs in Optimization, DOI 10.1007/978-1-4939-2282-6_8

Fig. 8.1 L-shaped method
decomposition strategy [7]

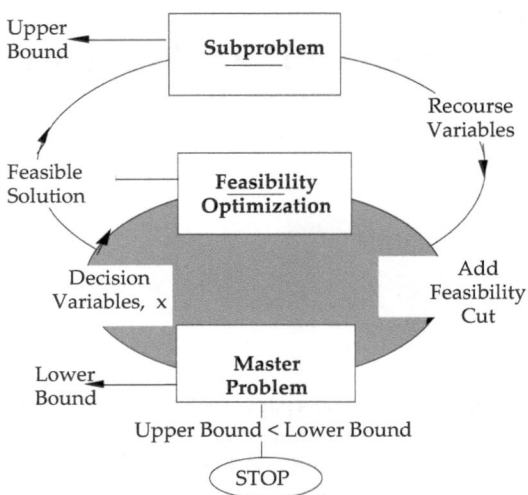

addition is continued until no constraint is violated (completely feasible solution). It should be noted that this is a very time-consuming iterative loop of the L-shaped algorithm, and variants of the L-shaped method provide improvements to this loop. The master problem then provides the values of the action variables x, and the lower bound to the objective function. At each outer iteration, for these fixed x, the sub-problem is solved for r, and linearizations of the objective and recourse function (optimality cuts) are obtained along with the values of r. If the subproblem solution (upper bound) crosses or is equal to the lower bound predicted by the master problem, then the procedure stops, else iterations continue.

Feasibility Optimization

$$\text{Minimize } \textit{Constraints Violations}(x^*, r) \qquad\qquad (8.4)$$
$$r, \lambda$$

The following example uses the news vendor problem to show the convergence of the L-shaped method. As indicated earlier, the inner loop of the L-shaped method consists of determining whether a first-stage decision is also second-stage feasible, and so on. This step is extremely computationally intensive and may involve several iterations per constraint for successive candidate first-stage solutions. In some cases though (such as this news vendor problem) this step can be simplified. A first case is when the second-stage is always feasible. The stochastic program is then said to have *complete recourse* .

Example 8.1 The simplest form of a stochastic program may be the news vendor (also known as the newsboy) problem. In the news vendor problem, the vendor must

Table 8.1 Weekly demand uncertainties

j	Demand, d_j	Probability, p_j
1	50	5/7
2	100	1/7
3	140	1/7

determine how many papers (x) to buy now at the cost of c cents for a demand which is uncertain. The selling price is s_p cents per paper. For a specific problem, whose weekly demand is shown below, the cost of each paper is $c = 20$ cents and the selling price is $s_p = 25$ cents. Solve the problem, if the news vendor knows the demand uncertainties (Table 8.1) but does not know the demand curve for the coming week a priori. Assume no salvage value $(s = 0)$, so that any papers bought in excess of demand are simply discarded with no return.

Solve this problem using the L-shaped method.

Solution The formulation of the problem is given below.

$$\text{Maximize} \quad -Z = Profit_{avg}(u) \tag{8.5}$$
$$x$$

$$Profit_{avg}(u) = \int_0^1 [-cx + Sales_p(r, w, p(u))]dp$$

$$= \sum_j p_j Sales(r, w, d_j) - cx \tag{8.6}$$

$$Sales(r, w, d_j) = s_p r_j + s w_j \tag{8.7}$$

$$r_j = \min(x, d_j)$$
$$= x, \text{ if } x \leq d_j$$
$$= d_j, \text{ if } x \geq d_j \tag{8.8}$$
$$w_j = \max(x - d_j, 0) \tag{8.9}$$

where $Sales_p$ represents the recourse function R given below. We are minimizing Z or maximizing $-Z$.

$$R = s_p x$$
$$\text{if } 0 \leq x \leq d_1 \tag{8.10}$$

or

$$R = 5/7 s_p d_1 + 1/7 s_p x + 1/7 s_p x$$
$$\text{if } d_1 \leq x \leq d_2 \tag{8.11}$$

or

$$R = 5/7 s_p \, d_1 + 1/7 s_p \, d_2 + 1/7 s_p \, x$$
$$\text{if} \quad d_2 \leq x \leq d_3 \tag{8.12}$$

As can be seen from the above formulation, this problem does not have any equality terms and is considered a problem with complete recourse. To obtain the optimal solution, we need to consider the outer loop iterations (no feasibility cut) given in Fig. 8.1 for the L-shaped method. From Table 8.1, we know that the uncertain parameter u can take values 50, 100, 140, with probabilities 5/7, 1/7, and 1/7, respectively. The recourse function can be calculated using analytical expressions given in Eqs. 8.10–8.12 and hence sampling can be avoided. Figure 8.2 shows the terms in the recourse function $Sales_p(50)$ and $Sales_p(100)$. Each of these functions is polyhedral. The sequence of iterations for the L-shaped method is given below.

1. Assume $x = 100$ and assume the lower bound to be $-\infty$. The recourse function that is calculated by the subproblem is calculated using Eqs. (8.6)–(8.9) and is equal to $Profit = -393$. To express this in the minimization term, $Z_{up} = 393$.
2. The linear cut (Eq. (8.14)) for the recourse function derived from Eq. (8.11) is added to the master problem, given below.

$$\text{Maximize} \quad -Z_{lo} = -20x + R \tag{8.13}$$
$$x$$

$$R \leq 25 \left(\frac{5}{7} \times 50 + \frac{2}{7} \times x \right) \quad \text{linear cut at } x = 100 \tag{8.14}$$

The solution to the above problem is $x = 0$ and $Z_{lo} = -892.86$. The recourse function calculated again using the Eqs. (8.6)–(8.9) is equal to $Z_{up} = 0$. The solution is not optimal as the upper bound (0) is greater than the lower bound (-892.86).
3. Add a new cut, Eq. (8.17), and solve the following problem.

$$\text{Maximize} \quad -Z_{lo} = -20x + R \tag{8.15}$$
$$x$$

$$R \leq 25 \left(\frac{5}{7} \times 50 + \frac{2}{7} \times x \right) \text{ linear cut at } x = 100 \tag{8.16}$$

$$R \leq 25x \text{ linear cut at } x = 0 \text{ from Eq. (8.10)} \tag{8.17}$$

The solution to the above problem is $x = 50$ and $Z_{lo} = -250$. The recourse function at $x = 50$ is equal to $Z_{up} = -250$, and is the optimum. So the average profit per day is 250 cents with a total weekly profit of 1750 cents, as found before.

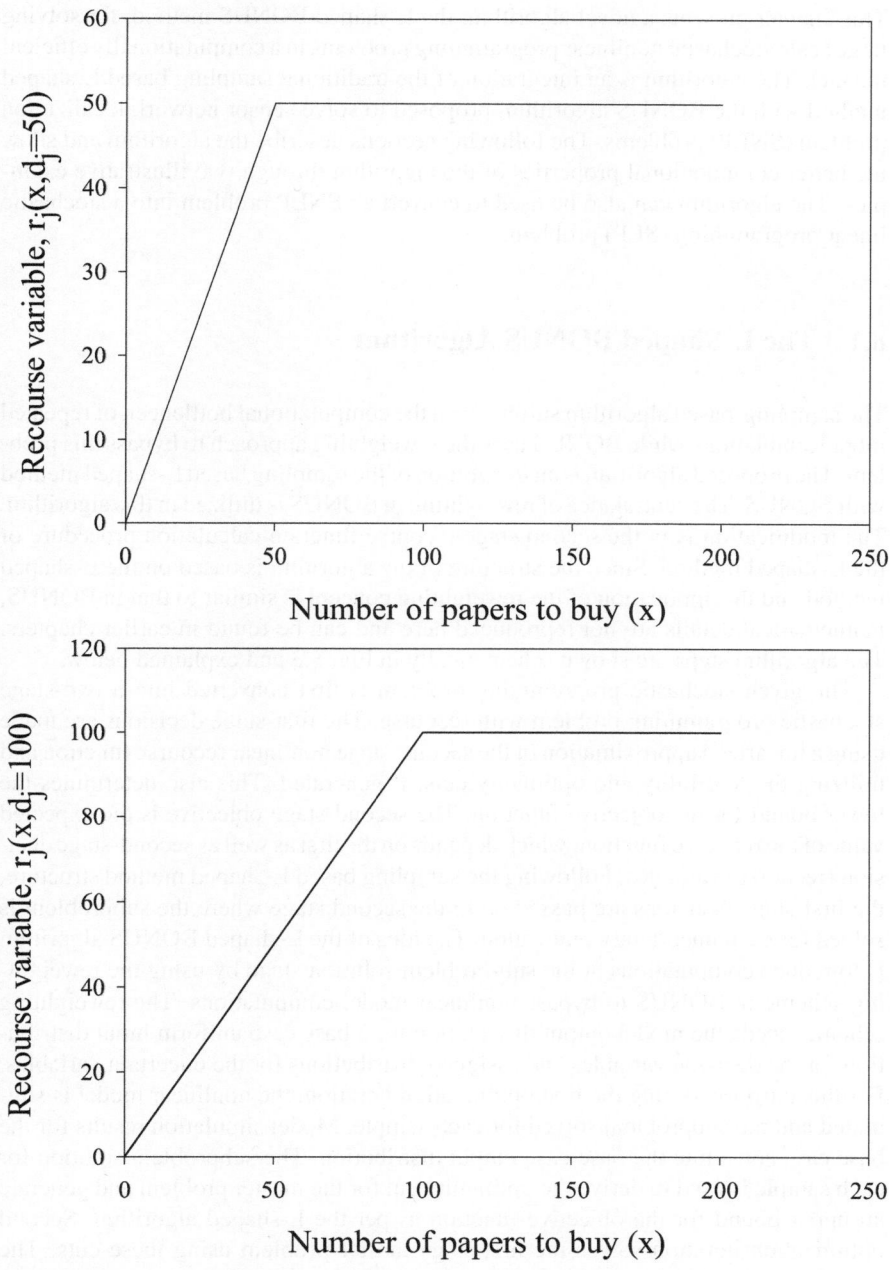

Fig. 8.2 Recourse function term as a function of the decision variable

This chapter presents a novel algorithm, the L-shaped BONUS method, for solving large scale stochastic nonlinear programming problems in a computationally efficient manner. The algorithm is an integration of the traditional sampling based L-shaped method with the BONUS algorithm, proposed to solve sensor network localization problem (SNLP) problems. The following sections describe the algorithm and show the better computational properties of the algorithm through two illustrative examples. The algorithm can also be used to convert an SNLP problem into a stochastic linear programming (SLP) problem.

8.1   The L-Shaped BONUS Algorithm

The sampling-based algorithm suffers from the computational bottleneck of repeated model simulations while BONUS uses the reweighting approach to bypass this problem. The proposed algorithm is an integration of the sampling based L-shaped method with BONUS. The central idea of reweighting in BONUS is utilized in this algorithm. The modification is in the second stage recourse function calculation procedure of the L-shaped method. Since the structure of the algorithm is based on the L-shaped method and the application of the reweighting concept is similar to that in BONUS, mathematical details are not reproduced here and can be found in earlier chapters. The algorithm steps are shown schematically in Fig. 8.3 and explained below.

The given stochastic programming problem is first converted into a two-stage stochastic programming problem with recourse. The first-stage decisions are made using a linearized approximation of the second-stage nonlinear recourse function and utilizing the feasibility and optimality cuts, if generated. This also determines the lower bound for the objective function. The second-stage objective is the expected value of the recourse function, which depends on the first as well as second-stage decision (recourse) variables. Following the sampling based L-shaped method structure, the first-stage decisions are passed on to the second stage where the subproblem is solved for each uncertainty realization. The idea of the L-shaped BONUS algorithm is to reduce computations at the sub-problem solution stage by using the reweighting scheme in BONUS to bypass nonlinear model computations. The reweighting scheme, needs the model output distribution for a base case uniform input distribution for the decision variables and assigned distributions for the uncertain variables. For this purpose, during the first optimization iteration, the nonlinear model is simulated and the subproblem solved for each sample. Model simulation results for the base case constitute the base case output distribution. The subproblem solution for each sample is used to derive the optimality cut for the master problem and generate an upper bound for the objective function as per the L-shaped algorithm. Second optimization iteration solves the first-stage master problem using these cuts. The new first-stage decisions along with an updated lower bound are passed on to the second-stage problem. During this iteration, when a new set of samples are taken by the stochastic modeler, model simulation, and the subproblem optimization solution is not performed for each sample. The reweighting scheme, with Gaussian kernel

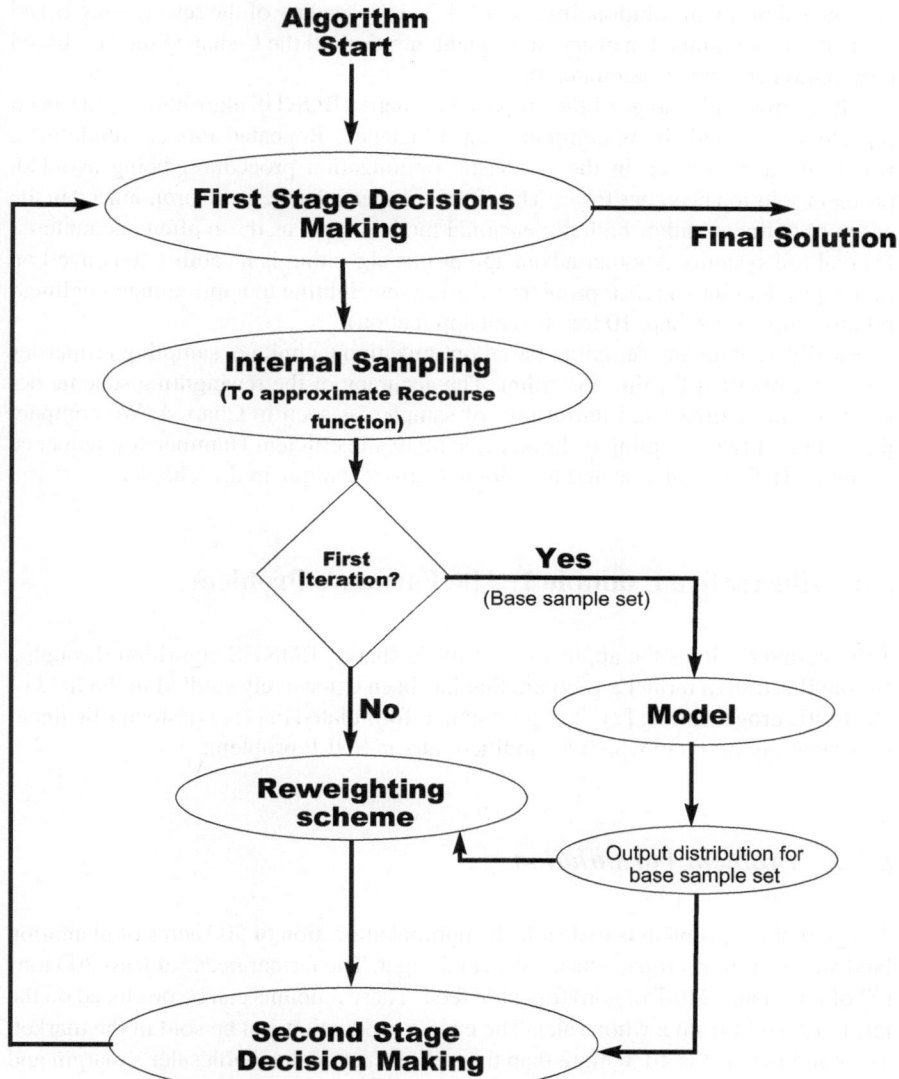

Fig. 8.3 The proposed L-shaped BONUS algorithm structure

density estimation, is used to predict the probabilistic values (expectations) of the model output. The base case output distribution along with the two sample sets are used for this prediction. The expected value of the model output is used to solve only one second-stage dual subproblem to generate cuts and update the objective function upper bound. It should thus be noted that for second iteration, only one subproblem is solved. Therefore, not only the nonlinear model simulation time but also the subproblem solution time is saved. This procedure of the reweighting based estimation is continued in every subsequent iteration till the L-shaped method based termination criteria is encountered.

The primary advantage of the proposed L-shaped BONUS algorithm, as has been repeatedly stressed, is its computational efficiency. Repeated model simulations, which are a bottleneck in the stochastic optimization procedure, being avoided, problem solution becomes faster. The effect is expected to be more pronounced in the case of nonlinear and/or high dimensional models, such as those often encountered for real life systems. Another advantage of this algorithm is its ability to convert an SNLP problem into an SLP problem by using reweighting to approximate nonlinear relationships (see Chap. 10 for such an application).

Finally, as with any sampling based optimization technique, sampling properties are very important for this algorithm. The accuracy of the reweighting scheme depends on the number and uniformity of samples as seen in Chap. 4. We compare the results of two sampling techniques, namely, the efficient Hammersley sequence sampling (HSS) technique and the Monte Carlo technique in this chapter.

8.2   Illustrative Example 1: The Farmer's Problem

This section explains the application of the L-shaped BONUS algorithm through a simple illustrative farmer's problem that has been extensively studied in the field of stochastic programming [2] . The problem, as formulated in [2] , is a stochastic linear programming problem which is modified into an SNLP problem.

8.2.1   Problem Formulation

The goal of the problem is to decide the optimal allocation of 500 acres of plantation land amongst three crops: wheat, corn, and sugar. The farmer needs at least 200 tons (T) of wheat and 240 T of corn for cattle feed. These amounts can be produced on the farm or bought from a wholesaler. The excess production can be sold in the market. The purchase cost is 40 % more than the selling cost due to wholesaler's margin and transportation cost. Sugar beet sells at a cost of $ 36/T if the amount is less than 6000 T. Any additional quantity can be sold at only $ 10/T. Through experience, the farmer knows that the mean yield of crops is 2.5 T, 3 T, and 20 T per acre for wheat, corn and sugar, respectively. But these values are uncertain owing to various factors. The objective is to maximize the expected profit in the presence of uncertain yields. Table 8.2 summarizes the data and more details about the SLP can be found in [2] .

For this illustration, to convert the problem into an SNLP problem, the uncertain yield is assumed to be dependent on four different factors which are uncertain. These four factors are the average rainfall, availability of sunlight, attack probability of a crop disease, and the probability of attack by pests. The annual yield of the crops is

Table 8.2 Data for farmer's problem

	Wheat	Corn	Sugar beets
Yield maximum (T/acre)	3	3.6	24
Planting cost ($/acre)	150	230	260
Selling price ($/T)	170	150	36 under 6000 T
			10 above 6000 T
Purchase price ($/T)	238	210	–
Minimum requirement (T)	200	210	–

nonlinearly related to these four factors. Although the relationships presented here are hypothetical and simplistic, it is expected that some nonlinear equations will govern these relationships. The dependencies are as follows:

$$Y_r = 2\,\alpha_r \left(1 - \frac{\alpha_r}{2}\right) \qquad \alpha_r = Uniform[0.75, 1.25] \qquad (8.18)$$

$$Y_s = 1.58\,(1 - e^{-\alpha_s}) \qquad \alpha_s = Uniform[0.9, 1] \qquad (8.19)$$

$$Y_d = 1 - \alpha_d \qquad \alpha_d = Uniform[0, 0.2] \qquad (8.20)$$

$$Y_p = 1 - \alpha_p^2 \qquad \alpha_p = Uniform[0, 0.2] \qquad (8.21)$$

where

- Y_j are fractions of the maximum yield due to corresponding effects j
- α_r: Fractional rainfall of the yearly average (uniform distribution between 0.75 and 1.25)
- α_s: Fractional sunlight of the yearly average (uniform distribution between 0.9 and 1)
- α_d: Attack probability of a crop disease (uniform distribution between 0 and 0.2)
- α_p: Attack probability of pests (uniform distribution between 0 and 0.2)

The overall fractional yield of the crops is given by

$$YY_{i,actual} = Y_r \times Y_s \times Y_d \times Y_p \times YY_{i,max} \qquad (8.22)$$

where $YY_{i,actual}$ is the actual yield of the crop i and $YY_{i,max}$ is the maximum possible yield if all the conditions are perfect given in Table 8.2. Once these equations are incorporated in the original model, the resulting stochastic programming problem is given as:

Minimize $150x_1 + 230x_2 + 260x_3$

$$+ E[238y_1 - 170w_1 + 210y_2 - 150w_2 - 36w_3 - 10w_4]$$

subject to the following constraints

$$x_1 + x_2 + x_3 \leq 500,$$
$$Y_{1,actual} x_1 + y_1 - w_1 \geq 200,$$
$$Y_{2,actual} x_2 + y_2 - w_2 \geq 240,$$
$$w_3 + w_4 \leq Y_{3,actual} x_3,$$
$$w_3 \leq 6000,$$
$$x_1, x_2, x_3, y_1, y_2, w_1, w_2, w_3, w_4 \geq 0,$$

where E is the expectation operator over the uncertain variables $YY_{i,actual}$, x are decision variables related to land assignments, w are sales, and y are purchases. $YY_{i,actual}$ is the yield of crop i given by Eq. (8.22) and *nonlinearly* related to the uncertain variables through Eqs. (8.18)–(8.21).

This problem when converted into a two-stage stochastic programming problem with recourse is given as:
First-Stage Problem

$$\text{Min} \quad 150x_1 + 230x_2 + 260x_3 + \theta$$
$$\text{s.t.} \quad x_1 + x_2 + x_3 \leq 500,$$
$$G_l\, x + \theta \geq g_l \quad l = 1 \ldots s,$$
$$x_1, x_2, x_3 \geq 0,$$

where θ is the linear approximation of the expected value of the recourse function. x_1, x_2, and x_3 constitute the first-stage decision variables. The constraints include the problem defined constraints on the first-stage decision variables and optimality cuts applied during iterations of the L-shaped method.
Second-Stage Problem

$$Q(x, \xi) = \min\{238y_1 - 170w_1 + 210y_2 - 150w_2 - 36w_3 - 10w_4\}$$
$$\text{s.t.} \quad Y_{1,actual} x_1 + y_1 - w_1 \geq 200,$$
$$Y_{2,actual} x_2 + y_2 - w_2 \geq 240,$$
$$w_3 + w_4 \leq Y_{3,actual} x_3,$$
$$w_3 \leq 6000,$$
$$y_1, y_2, w_1, w_2, w_3, w_4 \geq 0.$$

Here, y_1, y_2, w_1, w_2, w_3, and w_4 are the second-stage decision variables (recourse variables). The constraints on the recourse variables in the original problem are considered in the second-stage problem solution. Note that in this problem, there are no equality constraints so this problem is also a problem with complete recourse like the news vendor problem presented earlier.

8.2.2   Problem Solution

The problem, when solved using sampling based L-shaped method, involves dual formulation of the nonlinear second-stage problem and solution of the dual problem in the second stage for each sample from the given sample set. Even if the nonlinearity is separated from the problem by considering directly the yield in the second-stage problem (in place of the nonlinear relationships), the task of finding dual problem solutions for the samples can be demanding. The L-shaped BONUS algorithm can simplify the task by using reweighting to bypass the nonlinear model, as represented by Fig. 8.3. The ability of reweighting to effectively model the nonlinear relationship between the uncertain parameters and crop yield will help in converting the problem into an SLP with reduced computations.

The exact solution procedure is as follows.

- At every second-stage problem solution, uncertain parameters are sampled n times, n being a predecided sample size.
- During the first iteration, the samples are used to calculate the value of crop yield ($Y_{i,actual}$ and the yield value is used to solve the dual (as dual representation only depends on the crop yield) for each sample (i.e., n dual problem solutions) and an optimality cut, if needed, is generated.
- The first sample set is stored as the base sample set. At subsequent iterations, during the second-stage solution, the new set of n samples are taken and instead of solving the dual for each sample through yield calculation, reweighting is used to calculate the expected value of the crop yield.
- This single expected value is used in the dual problem which is now converted into a linear one. Moreover, with one expected value of the yield, the dual problem needs to be solved only once to calculate the expected value of the recourse function and generate the cut if needed. Use of reweighting therefore simplifies the problem on two counts. First, it bypasses the nonlinear part of the model and converts it into a linear model and second, computations are simplified by solving just one optimization subproblem at the second stage.

Reproduced below are the the first two iterations of the problem solution to explain the steps.

Solution: Iteration 1

- Step 0: $s = 0$ (iteration count)
- Step 1: $\theta^1 = -\infty$ (very low value). Solve

$$\text{Min} \quad 150x_1 + 230x_2 + 260x_3$$

$$\text{s.t.} \quad x_1 + x_2 + x_3 \leq 500$$

$$x_1, x_2, x_3 \geq 0$$

The solution is $x_1^1 = x_2^1 = x_3^1 = 0$
- Step 2: Sample the uncertain variables n times to generate the base sample set $\{u*\}$

- Step 3: Calculate the yield (n values) of the crop using the n sampled uncertain variables and Eqs. (8.18)–(8.21).
- Step 4: Solve the following dual problem for the n samples of crop yield. The values of x_i^1 are passed on to the second stage.

$$\text{Max} \quad \lambda_1(200 - Y_{1,actual}\ x_1^1) + \lambda_2(240 - Y_{2,actual}\ x_2^1) - \lambda_3(Y_{3,actual}\ x_3^1) - 6000\lambda_4$$

$$\text{s.t.} \qquad \lambda_1 \leq 238$$

$$\lambda_2 \leq 210$$

$$\lambda_1 \geq 170$$

$$\lambda_2 \geq 150$$

$$\lambda_3 + \lambda_4 \geq 36$$

$$\lambda_3 \geq 10$$

$$\lambda_1, \lambda_2, \lambda_3, \lambda_4 \geq 0 \qquad\qquad\qquad (8.23)$$

The solution of problem (8.23) for first sample is $\lambda_1 = 236$, $\lambda_2 = 210$, $\lambda_3 = 36$, $\lambda_4 = 0$. The expected value of the recourse function (w) calculated after all the dual problem solutions is $w = 98000$. Since $w > \theta$, an optimality cut is introduced.

Iteration 2

- Step 0: $s = 1$ (iteration count)
- Step 1: Solve

$$\text{Min} \qquad 150x_1 + 230x_2 + 260x_3 + \theta$$

$$\text{s.t.} \qquad x_1 + x_2 + x_3 \leq 500$$

$$\theta \geq 98000 - [610.1 \quad 636.4 \quad 727.4][x_1 \ x_2 \ x_3]^T$$

$$x_1, x_2, x_3 \geq 0 \qquad\qquad\qquad (8.24)$$

The solution of problem (8.24) is $x_1 = 0$, $x_2 = 0$, $x_3 = 500$ and $\theta = -264685.247$.
- Step 2: Sample the uncertain variables n times to generate the new sample set $\{u\}$
- Step 3: Calculate the estimated yield of the crops using the base and new sample sets bypassing relations (8.18) to (8.21). The estimated yield is 0.842.
- Solve the dual problem given by equation set (8.23) only once using the estimated value of the crop yield. The solution of the problem is $\lambda_1 = 238$, $\lambda_2 = 210$, $\lambda_3 = 10$, $\lambda_4 = 26$. The expected value of the recourse function (w) calculated after the dual problem solutions is $w = -158986.526$. Another optimality cut is introduced.

The procedure is then followed according to iteration 2 (using reweighting instead of n dual problem solutions) until the termination criteria of $w \leq \theta$ is satisfied.

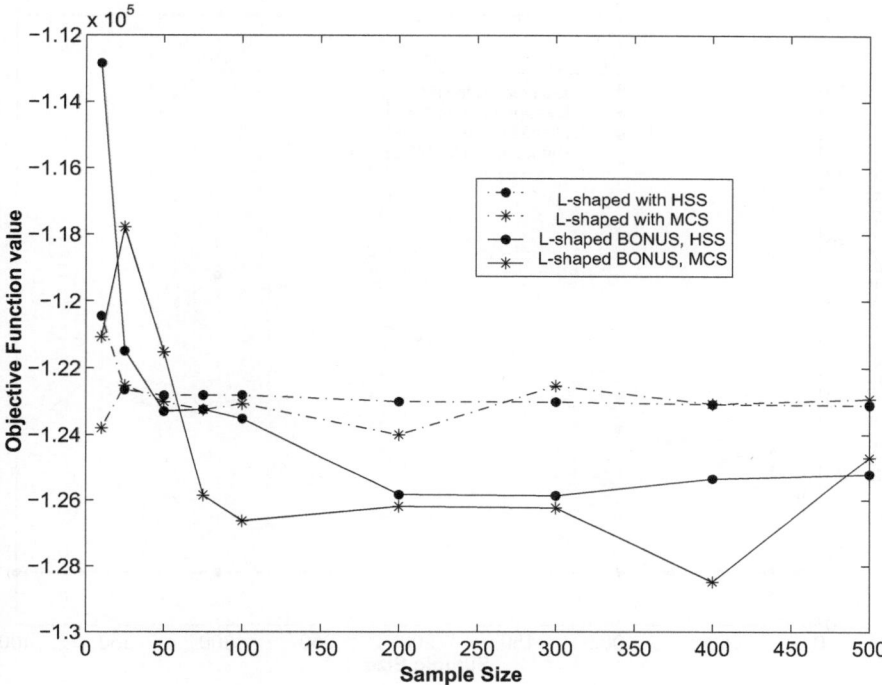

Fig. 8.4 Variation of objective function with sample size for farmer's problem

8.2.3   Results of the Farmer's Problem

Comparative results for the farmer's problem using the sampling based L-shaped method and the L-shaped BONUS algorithm are shown graphically in Fig. 8.4, which also compare the results for two different sampling techniques, Monte Carlo sampling (MCS) and HSS. Figure 8.4 compares the objective function values at the final solution as a function of the sample size. It is seen that the solutions for both algorithms approach a steady state value with increasing sample size. Moreover, the difference in the results for the two algorithms is within reasonable limits of sampling error, 1.7 % for the maximum sample size, indicating that reweighting approximation is not sacrificing accuracy. Based on this plot, HSS emerges as a more efficient sampling technique than MCS. The results for HSS appear to reach the steady state value faster than for MCS as the sample size is increased. This claim is further corroborated by Fig. 8.5 which plots the iterations needed to reach the solution for different sample sizes. It can be observed that for the standard L-shaped method, MCS sampling technique needs more iterations in general than HSS technique. The previously mentioned k dimensional uniformity property of HSS accounts for this observation. For the same sample size, the HSS method therefore approximates a given distribution better than the MCS. This results in faster convergence of HSS

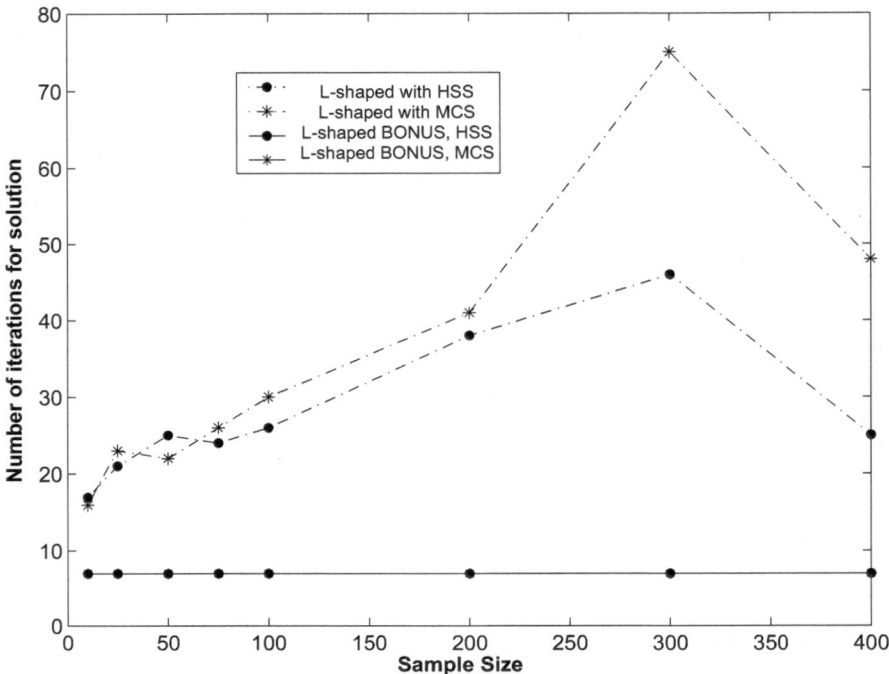

Fig. 8.5 Variation of iteration requirement with sample size for farmer's problem

based algorithms in general. For the L-shaped BONUS algorithm though, both MCS and HSS sampling techniques need six iterations irrespective of the sample size. This is possibly due to the approximation introduced by the reweighting scheme. The approximation renders the iteration requirements insensitive to sample size and sampling method changes. However, better values of final solutions (Fig. 8.4) confirm the superiority of the HSS method over MCS.

Computational time is an important factor while comparing these algorithms. Computational time increases exponentially with sample size for the standard L-shaped method while it increases almost linearly for the proposed L-shaped BONUS algorithm. The computational efficiency of L-shaped BONUS therefore becomes more pronounced as the sample size is increased. With the need to increase sample size to improve accuracy, the proposed algorithm offers a distinct advantage.

8.3   Illustrative Example 2: The Blending Problem

The problem reported here is typical for a petroleum industry manufacturing finished petroleum products such as lube oils. A large number of natural lubricating and specialty oils are produced by blending a small number of lubricating oil base stocks and additives. The lube oil base stocks are prepared from crude oils by distillation. The additives are chemicals used to give the base stocks desirable characteristics which they lack of to enhance and improve existing properties [13, 37] . In the context of such an application, a general chemical blending optimization problem is explained below followed by results comparing different solution and sampling techniques.

8.3.1 Problem Formulation

The aim is to blend n different chemicals (such as lube oil base stocks and additives) to form p different blend products (lube oils) at a minimum overall cost. Each chemical (base stock) has varying fractions of m different components (such as C_1–C_4 fraction, C_5–C_8 fraction, heavy fraction, inerts, etc.). Market demands call for production of a particular quantity of each blend product. Blend products catering to different applications (e.g. high performance lube oil, grease, industrial grade lube oil, etc.) have different specifications on fractions of m different components (for a lube oil such specifications will depend on physical property requirements like pour point, viscosity, boiling temperature). These specifications need to be satisfied to market the blend products. The task is complicated due to the presence of q impurities in the chemicals. Exact mass fractions of these impurities in some of the chemicals (base stocks) are uncertain. Such uncertainties may arise when the chemicals to be blended are themselves the product of other processes (such as crude distillation for lube oil base stocks). There are also specifications on the maximum amount of impurity in a blend product. If the impurity content of a blend product does not satisfy the regulation, the product has to be treated to reduce impurities to levels below specifications. The treatment cost depends on the amount of reduction in the impurities to be achieved. The goal in formulating the stochastic optimization problem is to find the optimum blend policy to minimize raw material cost and expected blend product treatment cost in the presence of uncertainty associated with impurity content of the chemicals. The stochastic programming problem is formulated as below.

$$\text{Minimize} \quad \sum_{i=1}^{n}\sum_{k=1}^{p} C_i W_{ik} + E\left[\sum_{k=1}^{p} C_T \theta_k\right] \tag{8.25}$$

Subject to:

$$\sum_{i=1}^{n} W_{ik} = \bar{W}_k \qquad \forall k = 1,\ldots,p \tag{8.26}$$

$$\sum_{i=1}^{n} x_{ij} W_{ik} \geq \bar{x}_{jk} \qquad \forall k = 1,\ldots,p \quad \text{and} \quad j = 1,\ldots,m \tag{8.27}$$

$$\sum_{l=1}^{q} (I_{il}(u))^{\alpha_l} = I_i^* \qquad \forall i = 1,\ldots,n \tag{8.28}$$

$$\sum_{i=1}^{n} I_i^* W_{ik} = \bar{I}_k \qquad \forall k = 1, \ldots, p \tag{8.29}$$

$$\bar{I}_k . (1 - \theta_k) \leq I_k^{spec} \qquad \forall k = 1, \ldots, p \tag{8.30}$$

Here, W_{ik} is the weight of chemical i in blend product k. C_i is the per unit cost of chemical i, while C_T is the blend product treatment cost per unit reduction in the impurity content. $\bar{W}_k$ is the total production requirement of blend product k. x_{ij} is the fraction of component j in chemical i and $\bar{x}_{jk}$ is the specification of component j in blend product k. $I_{il}(u)$ is the (possibly uncertain) fraction of impurity l in chemical i and I_i^* is the "impurity parameter" of chemical i. This impurity parameter gives the extent to which a chemical is impure, as a nonlinear function of various impurities. Coefficients α_l decide the importance of a particular impurity in the final product. $\bar{I}_k$ is the final impurity parameter of a blend which depends on the weight contribution of each chemical in a particular blend. I_k^{spec} is the maximum permitted impurity content in the blend product. θ_k is the purification required for blend k to satisfy the impurity constraint.

The objective function consists of two parts. The first part is the cost of chemicals used to manufacture the blend products and the second part is the expected treatment cost of the off-spec products. The first set of constraints ensures the required production of each blend product. The second constraint set ensures that component specifications for the blended products are satisfied. These specifications are expressed in terms of the minimum amount of each component needed in the blend product. The third set of constraints calculates the impurity parameter for each chemical, as a function of various individual impurities. The fourth equation calculates the "impurity parameter" for each blend product depending on the blending policy. The last set of constraints makes sure that all the impurity related specifications are satisfied by each blend product.

In sampling-based algorithms, the expected cost is calculated using various realizations of uncertain parameters (i.e., samples) and the corresponding treatment costs. Parameter $I_{il}(u)$ is then a function of each sample. The two-stage stochastic programming blending problem is given as

First-stage problem

$$\text{Minimize} \qquad \sum_{i=1}^{n} \sum_{k=1}^{p} C_i W_{ik} + E[R(W, \theta, u)] \tag{8.31}$$

where

$$\sum_{i=1}^{n} W_{ik} = \bar{W}_k \qquad \forall k = 1, \ldots, p \tag{8.32}$$

$$\sum_{i=1}^{n} x_{ij} W_{ik} \geq \bar{x}_{jk} \qquad \forall k = 1, \ldots, p \quad \text{and} \quad j = 1, \ldots, m \tag{8.33}$$

Table 8.3 Data for chemicals in blending problem

	A_1	A_2	A_3	A_4	A_5	A_6	A_7
C_1 fraction	0.20	0.10	0.50	0.75	0.10	0.30	0.20
C_2 fraction	0.10	0.15	0.20	0.05	0.70	0.30	0.55
C_3 fraction	0.60	0.65	0.22	0.12	0.10	0.30	0.16
I_1 fraction	0.02	0.07	0.01	0.02	0.043	0.015	0.012
I_2 fraction	0.01	0.005	0.02	0.02	0.01	0.04	0.021
I_3 fraction	0.06	0.023	0.02	0.03	0.022	0.028	0.055
Cost (\$/unit weight)	104	90	135	130	115	126	120

Here $E[R(W, \theta, u)]$ is the expected value of the recourse function which is calculated in the second stage.

Second-stage problem

$$\text{Minimize} \qquad E[R(W, \theta, u)] = \sum_{r=1}^{N_{samp}} \sum_{k=1}^{p} C_T \theta_k \qquad (8.34)$$

where

$$\sum_{l=1}^{q} (I_{il}(r))^{\alpha_l} = I_i^* \qquad \forall i = 1, \ldots, n \qquad (8.35)$$

$$\sum_{i=1}^{n} I_i^* W_{ik} = \bar{I}_k \qquad \forall k = 1, \ldots, p \qquad (8.36)$$

$$\bar{I}_k.(1 - \theta_k) \leq I_k^{spec} \qquad \forall k = 1, \ldots, p \qquad (8.37)$$

The first-stage decision variables are W_{ik}. The second stage considers various realizations of uncertain parameters I_{il} through N_{samp} samples. This second-stage problem minimizes the expected value of the recourse function through decision variables θ_k. There is no equality constraint in the second stage so this again is a problem of complete recourse. No feasibility cut optimization is needed. This is a stochastic programming problem with a nonlinear relationship between second stage parameters I_{il} and I_i^*.

8.3.2   Simulations and Results

This work considers the problem with 7 chemicals $(A_1, \ldots, A_7)$, three components $(C_1, \ldots, C_3)$, three blend products $(P_1, \ldots, P_3)$, and three different impurities, such as sulfur, ash, and heavy residue, i.e., $n = 7$, $m = 3$, $p = 3$, and $q = 3$. Data for the problem is reported in Tables 8.3 and 8.4. α_1, α_2 and α_3 are 0.9, 1.3, and 1.4, respectively, and the purification cost C_T is \$ 10,000 per unit reduction in impurity.

Table 8.4 Data for blend
products

	P_1	P_2	P_3
C_1 fraction	0.1	0.6	0.2
C_2 fraction	0.5	0.1	0.1
C_3 fraction	0.2	0.2	0.5
Production (weight units)	100	120	130
I_k^{spec}	0.9	1.05	1.2

Each chemical has one uncertain impurity fraction. Here I_{12}, I_{15}, I_{23}, I_{26}, I_{27}, and I_{34} are uncertain, varying by $\pm 25\%$ around the values reported in Table 8.3. All uncertain parameters are normally distributed in the given range, i.e., $\pm 25\%$ range corresponds with the $\pm 3\sigma$ range where σ is the standard deviation of the normal distribution.

The problem is solved using the standard L-shaped algorithm and the proposed L-shaped BONUS algorithm, both using the HSS and MCS techniques. In the L-shaped BONUS algorithm, the nonlinear relationship between I_{il} and I_i^* is bypassed using reweighting scheme.

The optimum objective function value is plotted in Fig. 8.6 for different sample sizes. The results show that with the HSS technique, the average difference in the absolute values of the final objective function for the standard L-shaped and L-shaped BONUS algorithm is only 1.6%, and this difference reduces with increasing sample size. Figure 8.7 plots the number of iterations required to achieve the solution. It can be observed that the L-shaped algorithm consistently requires more iterations. It is also observed that the L-shaped BONUS algorithm achieves an average reduction of 75% in solution time over the standard L-shaped algorithm. This significant reduction accompanied by a relatively small change in the final results makes L-shaped BONUS algorithm very attractive. A comparison between the HSS and MCS techniques shows observations and conclusions similar to those for the farmer's problem. Thus the MCS technique in general requires more iterations than the HSS technique and results with HSS settle much faster than with MCS.

8.4   Summary

This chapter presented the L-shaped BONUS algorithm for two-stage stochastic programming problems. This algorithm exploits the structure of the problem as in L-shaped method and the efficiency of the reweighting scheme in BONUS for evaluating recourse function. Two illustrative examples, namely, the farmer's problem and the blending problem, are presented. The L-shaped BONUS method is not only computationally efficient but the computational efforts do not scale exponentially with number of samples as in the traditional L-shaped method but scale linearly. The next two chapters present real world applications of the L-shaped BONUS algorithm to large scale real world problems.

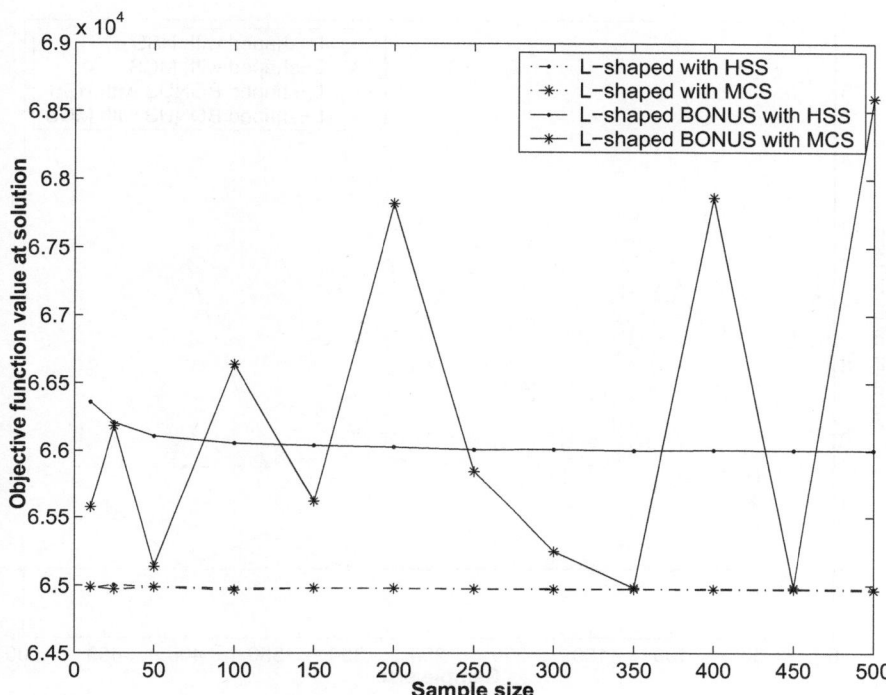

Fig. 8.6 Variation of objective function with sample size for blending problem

Notations

cdf_{out}	cumulative probability density function of output
C_i	per unit cost of chemical i
C_T	blend product treatment cost per unit reduction in the impurity content
d	demand
E	expected value function
f	a function
g	inequality constraint function
h	equality constraint function
$I_{il}(u)$	fraction of impurity l in chemical i
I_i^*	impurity parameter of chemical i
I_k	final impurity parameter of a blend which
	depends on the weight contribution of each chemical in a particular blend
I_k^{spec}	maximum permitted impurity content in the blend product
N_{samp}	number of samples
p	probability values
$P_i()$	probabilistic function

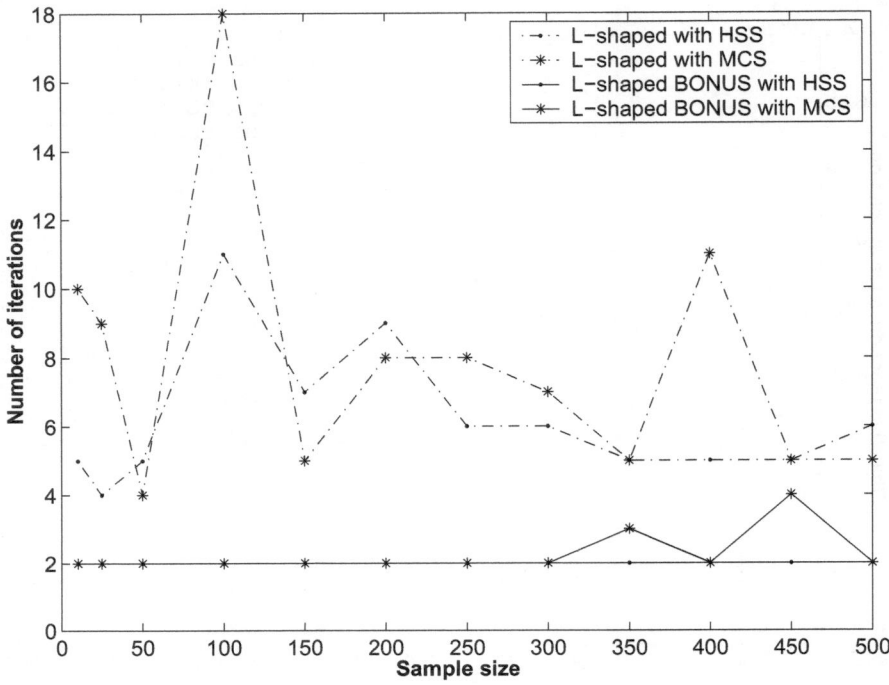

Fig. 8.7 Variation of iteration requirement with sample size for blending problem

$pdf_{in}()$	probability density function of input
$Q()$	recourse function in farmer's problem
r	recourse variable
$R()$	recourse function
s_p	selling price
w_i	amount of plant i sold
u	uncertain variable
W_{ik}	weight of chemical i in blend product k
$\bar{W}_k$	total production requirement of blend product k
x	decision variables
x_j	planting cost of crop j
x_{ij}	fraction of component j in chemical i
$\bar{x}_{jk}$	specification of component j in blend product k
w_j	sales cost of crop j
y_j	purchase cost of crop j
$YY_{i,actual}$	actual yield of the crop i

$YY_{i,max}$	maximum possible yield if all the conditions are perfect
Y_j	fractions of the maximum yield due to corresponding effects j
Z, z	objective function

Greek letters

α_d	Attack probability of a crop disease
	(uniform distribution between 0 and 0.2)
α_l	importance of a particular impurity in the final product
α_r	fractional rainfall of the yearly average
α_s	fractional sunlight of the yearly average
θ_k	purification required for blend k to satisfy the impurity constraint
λ	Lagrange multipliers/dual variables
σ	standard deviation

Chapter 9
The Environmental Trading Problem

9.1 Introduction

The increasing stringency of environmental regulations and the global rise of concerns about the environmental impact of industrial production have led to an increased focus on waste management decisions as a component of industrial sustainability. Pollutant credit trading, an approach that provides economic incentives for reducing pollution, is one novel idea introduced in an attempt to reduce the financial burden of waste management [56] . Both the US Environmental Protection Agency (USEPA) and the US Department of Agriculture (USDA) seek to promote this type of market-based solution. However, industry-level decision making under a pollutant trading scheme faces many difficulties, especially in the presence of uncertainty. In this chapter, the L-shaped BONUS algorithm is applied to the pollutant trading problem to optimize such decisions. This chapter is based on the paper by Shastri and Diwekar [51].

9.2 Basics of Pollutant Trading

Pollutant trading is a market-based approach to pollution reduction based on the idea that different pollutant sources have different pollution control costs. Therefore, if the task of pollutant reduction can be assigned to the facilities with the lowest control costs, the total cost of pollution control across all pollutant sources is minimized. A market-based trade mechanism accomplishes this by efficiently allocating pollutant reduction efforts among sources. With such a system in place, a facility that would otherwise exceed its allowable pollutant discharge has two options for meeting its regulatory obligations: (1) reducing its pollutant level or (2) paying another facility to reduce *its* pollutant level by an equivalent amount. The pollutant trading system therefore allows the facility to choose the lower cost option, whereas in the absence of such a system, it would have to reduce its own pollutant level at any cost.

In the case of watershed based trading, the total amount of pollutants that may be released into a watershed over time, while allowing the watershed to still meet

© Urmila Diwekar, Amy David 2015

U. Diwekar, A. David, *BONUS Algorithm for Large Scale Stochastic Nonlinear Programming Problems*, SpringerBriefs in Optimization, DOI 10.1007/978-1-4939-2282-6_9

water quality standards, is evaluated by the state or federal authority. This amount, in combination with factors such as the existing discharge levels, the expected discharge from nonindustrial sources, and the relative size and economic contributions of each polluter, leads to the establishment of a regulation such as a total maximum daily load (TMDL) for each polluter. A polluter may then bring itself into compliance by applying waste treatment methods, incurring both capital and operating costs that depend on the type and amount of waste treated, the existing technology, and the level of reduction to be achieved. If this is found to be a high-cost proposition, the facility may instead buy "credits" from another polluter, allowing it to release an amount of discharge greater than its TMDL into the watershed. The polluter that has sold credits, meanwhile, must reduce its discharge an equivalent amount below its TMDL [57].

The credit trading market is affected by transaction costs, number of participants, availability of cost data, and uncertainties related to continued industry participation and data availability. Trading economics are also influenced by the trading ratio, how many units of pollutant reduction a source must purchase to receive credit for one unit of load reduction.

Among the participants in the trading market, each polluter is classified as a point source (PS) or a nonpoint source (NPS). Point sources are those that have direct and measurable emissions, such as industries, while nonpoint sources have diffused emissions that are more difficult to measure, such as agricultural runoffs. Because nonpoint sources are the primary polluters in most watersheds, from a volumetric perspective, and because pollution control costs are typically lower for nonpoint sources, trading between point and nonpoint sources has considerable potential for pollution reduction.

However, the nonpoint sources are difficult to measure at a reasonable cost, and the diffusion of the pollutants makes it difficult to estimate the effectiveness of pollution reduction strategies. Further, pollution from nonpoint sources is often dependent on stochastic factors such as rainfall and other weather conditions. These factors introduce a significant amount of uncertainty into the economics of trading between point and nonpoint sources. Thus, in the presence of multiple polluters in both point and nonpoint categories, heuristics-based decision making is likely to be suboptimal. A framework based on mathematical modeling concepts and making use of the BONUS method can have significant value to industries in analyzing their options.

9.3   Christina Watershed Nutrient Management

The Christina watershed is an important watershed in the Lower Delaware River (LDR), draining three states and providing up to 100 million gallons of public drinking water per day [48]. The LDR Basin has had ongoing problems with both point and nonpoint sources of pollution, consisting primarily of industrial discharges, urban runoff, and agriculture. This has led to concern over both sediment and nutrients released into the watershed. Sediment consists of loose sand, clay, and other soil particles caused primarily by soil erosion and decomposition of plants and animals, and can be greatly accelerated by human use of land. Nutrients consist of nitrogen and phosphorus that find their way into the watershed through agricultural, storm water, wastewater, household, and industrial runoff [24].

Various watersheds within the LDR basin have been declared "impaired," having pollutant levels that exceed those allowable for maintenance of water quality. Thirty-nine segments of the basin have been declared impaired due to their low levels of dissolved oxygen (DO) and nutrient additions from various point and nonpoint sources, including industrial and municipal point sources, and agricultural, superfund, and hydromodification nonpoint sources are considered to be the major cause. Authorities have proposed two different TMDLs for the Christina watershed, a low-flow TMDL focusing on the impact of nitrogen and phosphorus additions from the point sources, and a high-flow TMDL accounting for the nonpoint source additions of bacteria and sediment. Trading has been proposed as a viable option to achieve reductions in phosphorus, total nitrogen, ammonia nitrogen, and carbonaceous biochemical oxygen demand (CBOD), allowing the point source load allocations to meet the low-flow TMDL targets.

The simplest trading mechanism would involve trading among the various point sources. However, opportunities for pollutant reduction in this manner are limited by similar treatment costs throughout the watershed. A more effective trading mechanism would instead leverage trading between point and nonpoint sources, as nonpoint sources offer significant opportunity to reduce pollution by converting existing agricultural land to forest or implementing best management practices (BMP) on the cultivated lands. Therefore, this chapter proposes a trading mechanism by which land is allocated to a particular point source and the point source is responsible for management of the nonpoint source to offset nutrient discharges from the point source facility.

9.4   Trading Problem Formulation

A general trading problem applicable to any watershed is formulated, and subsequently applied to the Christina watershed case study. Because of the uncertainity associated with the nonpoint sources, a stochastic programming problem is required. The problem considers trading between a set of point sources and a single nonpoint source, assumed to be a farm. All sources discharge pollutants into a common body of water, such as a lake. The maximum amount of discharge per day into the body of water is statutorily regulated. The model does not consider regulations on nonpoint source emissions, a simplification that reflects actual regulations that exclude nonpoint sources due to the impossibility of accurately measuring their emissions. The development of TMDL results, therefore, in specific load allocations for each point source that becomes the baseline for trading between the point source and the nonpoint source. Note that the reduction techniques for the nonpoint source are nonlinearly dependent on the type and quantity of pollutant being treated.

Uncertainties in both inputs and outputs to the sources necessitate that the problem be formulated as a stochastic program. Let $i = 1, ..., P$ represent the set of point source and $j = 1, ..., M$ represent the chemicals that are regulated. Assume that the current pollutant discharge levels and the discharge reduction cost are known for each chemical at every point source.

Additional parameters characterizing each point source are:

- $D(i)$, the total volumetric discharge from point source i, expressed as volume/time
- $e_{p0}(i, j)$, the pretreatment discharge quantity of chemical j from point source i, expressed as mass/volume
- $c_p(i, j)$, the cost of treating chemical j at point source i, expressed as dollars/mass.

Some point sources have uncertainties in the measurement of their discharge quality and quantity, typically introduced when a point source treats incoming wastewater from a variety of sources, and resulting in both inputs and outputs that vary within certain limits. For example, a publicly owned water treatment plant (POWT) may treat sewage waste as well as water runoffs, the latter having a variable quantity and content. Therefore, only $c_p(i, j)$ is a deterministic parameter for all i and j; both $D(i)$ and $e_{p0}(i, j)$ contain uncertainty.

The nonpoint source is assumed to have a fixed amount of available land that can be divided among all point sources to implement treatment technologies (BMP). The nonpoint source is characterized by:

- L_{max}, the maximum amount of nonpoint source land available for trading, expressed as area
- e_{n0}, the pretreatment discharge quantity of chemical j from the nonpoint source, expressed as mass·area/time
- $c_n(j)$, the cost for the nonpoint source discharge reduction of chemical j, expressed as dollars/area
- $b_{NPS}(j)$, the nonpoint source abatement efficiency of chemical j
- $q_n(j) = e_{n0}(j)b_{NPS}(j)$, the abatement in nonpoint source discharge of chemical j, expressed as mass·area/time.

As previously mentioned, there are difficulties in measuring both the emissions and the reduction efficiencies for a particular technology at a nonpoint source, thus the actual reductions achieved by BMP are not precisely known. Therefore, e_{no} and b_{NPS} (and thus $q_n(j)$) contain uncertainty, while other parameters are assumed to be known for all i and j.

In addition to the waste load allocation for each point source, there are regulatory restrictions on the maximum amount of any chemical that can be discharged into the water body at a particular location. Enforcing this limit ensures that the implementation of pollutant trading does not result in the creation of localized points of high pollutant concentration known as "hot spots." Accordingly, the model includes $z_{red}(ij)$ representing the targeted reduction in discharge of chemical j by point source i (expressed as mass/time) and $z_{allowed}(j)$ representing the maximum permitted discharge of chemical j at any single location (expressed as mass/time).

Two decisions are to be made, with the goal of achieving the reduction targets at the lowest *total* cost:

1. How much end-of-pipe treatment reduction should be achieved at each point source?

2. How much land (NPS) should be allocated to each point source to achieve further reductions?

Accordingly, the decision variables in the model are $q_p(i, j)$, the discharge abatement of chemical j at point source i (expressed as mass/volume), and $L(i, j)$, the land allocated for trading by point source i to treat chemical j. The objective function is therefore

$$Min f_1(c_n, L) + E[f_2(D, c_p, q_p)], \tag{9.1}$$

where E is the expectation operator over the uncertain parameters and f_1 and f_2 are nonlinear functions of the respective variables. The first term in the objective represents the cost incurred due to trading through the allocation of land for each point source from the nonpoint source, while the second term represents the expected value of the total end-of-pipe treatment cost incurred by the point source to satisfy the regulations in the presence of uncertainty.

A feasible solution to this problem must meet the following constraints :

$$\sum_{i,j} L(i, j) \leq L_{max} \tag{9.2}$$

$$E\left[D(i, j)q_p(i, j) + L(i, j)q_n(j)\right] \geq z_{red}(i, j) \qquad \forall\, (i, j) \tag{9.3}$$

$$E\left[D(i, j)\left(e_{p0}(i, j) - q_p(i, j)\right)\right] \leq z_{allowed}(j) \qquad \forall\, (i, j) \tag{9.4}$$

$$0 \leq q_p(i, j) \leq e_{p0}(i, j) \qquad \forall\, (i, j) \tag{9.5}$$

$$q_n(j) = e_{n0}(j)b_{NPS}(j) \qquad \forall\, (i, j) \tag{9.6}$$

The first constraint (9.2) ensures that the total land allocated to all point sources does not exceed the amount of land available at the nonpoint sources. The second set of constraints (9.3) ensures that each point source achieves its individual reduction target for each chemical, with or without trading, while the third set (9.4) ensures that the emission of pollutant j at any location does not exceed the maximum allowable amount. The reduction of each chemical at each point source is restricted to values between zero and the initial concentration by the fourth set of constraints (9.5). Finally, the fifth set of constraints (9.6) models the effect of uncertainty on the problem by relating the pollutant reduction by the nonpoint source $(q_n(j))$ to the uncertain parameters $e_{n0}(j)$ and $b_{NPS}(j)$.

The problem can be made more tractable by converting it into a two-stage stochastic programming problem with recourse. The first-stage decisions are land allocations (the trading itself) between the point source and the nonpoint source, $L(i, j)$, and the second-stage decisions are the amounts of point source abatement, $q_p(i, j)$, achieved by end-of-pipe treatment. The two-stage formulation, including specific definitions of functions f_1 and f_2, is given as

$$First - stage\ problem \tag{9.7}$$

$$Min \sum_{i=1}^{P} \sum_{j=1}^{M} c_n(i, j)L(i, j)^{\alpha_j} + E\left[R(L, q_p, q_n, D)\right] \tag{9.8}$$

Table 9.1 Point source details for Christina River Basin

Point source	Total discharge (MGD[a])	Current discharge (kg/day) N	Current discharge (kg/day) P	Targeted (% reduction) N	Targeted (% reduction) P	Treatment (cost, $/kg) N	Treatment (cost, $/kg) P
1	0.4	30.30	30.30	0	13	15.6	5.2
2	1.028	233.63	38.94	26	26	14	4.9
3	7.5	568.18	568.18	25	0	10.9	3.8
4	3.85	291.66	299.76	0	68	12.7	4.2
5	0.6	68.18	45.45	10	10	15.4	5.1
6	1.1	125.0	313.72	34	83	14.4	5
7	0.72	5.45	5.45	5	5	18.3	5.4
8	0.7	171.06	26.51	69	0	15.4	5.12

[a] *MGD* Millions of gallons per day

$$subject\ to:$$

$$\sum_{i=1}^{P}\sum_{j=1}^{M} L(i,j) \leq L_{max}, \tag{9.9}$$

where R is the recourse function. The term α_j is a constant for chemical j that represents the nonlinear relationship between land allocation and pollutant reduction.

$$Second-stage\ problem \tag{9.10}$$

$$Min\ E\left[R(L,q_p,q_n,D)\right] = \sum_{n=1}^{N_{samp}}\sum_{i=1}^{P}\sum_{j=1}^{M} D(i,j,n)c_p(i,j)q_p(i,j,n) \tag{9.11}$$

$$subject\ to:$$

$$D(i,j,n)q_p(i,j) + L(i,j)q_n(j,n) \geq z_{red}(i,j,n) \qquad \forall\,(i,j,n) \tag{9.12}$$

$$D(i,j,n)\left[e_{p0}(i,j) - q_p(i,j,n)\right] \leq z_{allowed}(j) \qquad \forall\,(i,j,n) \tag{9.13}$$

$$0 \leq q_p(i,j) \leq e_{p0}(i,j) \qquad \forall\,(i,j) \tag{9.14}$$

$$q_n(j,n) = e_{n0}(j)b_{NPS}(j,n) \qquad \forall\,(j,n), \tag{9.15}$$

where N_{samp} is the sample size used to represent the uncertain space in the optimization algorithm, and n is a particular sample from that space.

For the Christina watershed, the authorities have recommended 8 point sources for trading, out of a total of 104 point sources. Two of the eight are private industries, while the other are municipal polluters. Two nutrients, nitrogen and phosphorus, are considered tradable commodities, and both have known TMDL-generated reduction targets at each point source. The total volumetric discharge, current discharge levels, and reduction targets for both pollutants are given in Table 9.1, along with the mean values for the cost of end-of-pipe treatment at each point source. Various parameters

Table 9.2 Point source details of NPS emission and treatment

	Nitrogen	Phosphorus
Mean value of emission quantity (kg per unit area per day)	20.2	30.5
Standard deviation in emission quantity	2.0	2.0
BMP cost ($ per unit area)	17.18	17.18
BMP nutrient reduction efficiency	0.50	0.39
Standard deviation in reduction efficiency	0.02	0.02

BMP best management practice

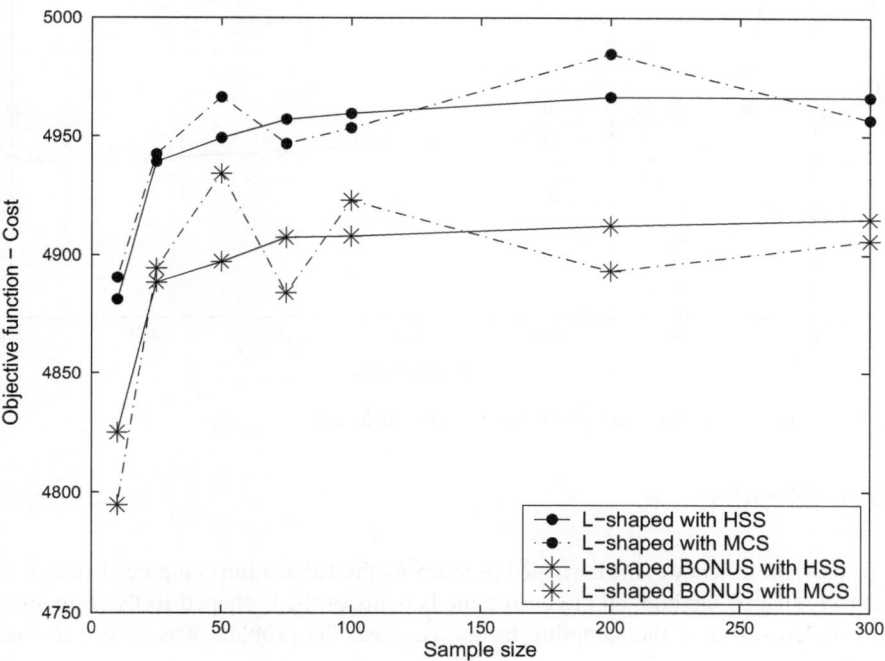

Fig. 9.1 Variation of the objective function result with sample size

of the BMPs along with the average pretrading discharges of both nutrients for the nonpoint source are given in Table 9.2. The maximum allowed concentrations at a discharge point ($z_{allowed}$) are 450 and 570 kg/day for nitrogen and phosphorus, respectively. The total quantity of land available for trading (L_{max}) is 500 units.

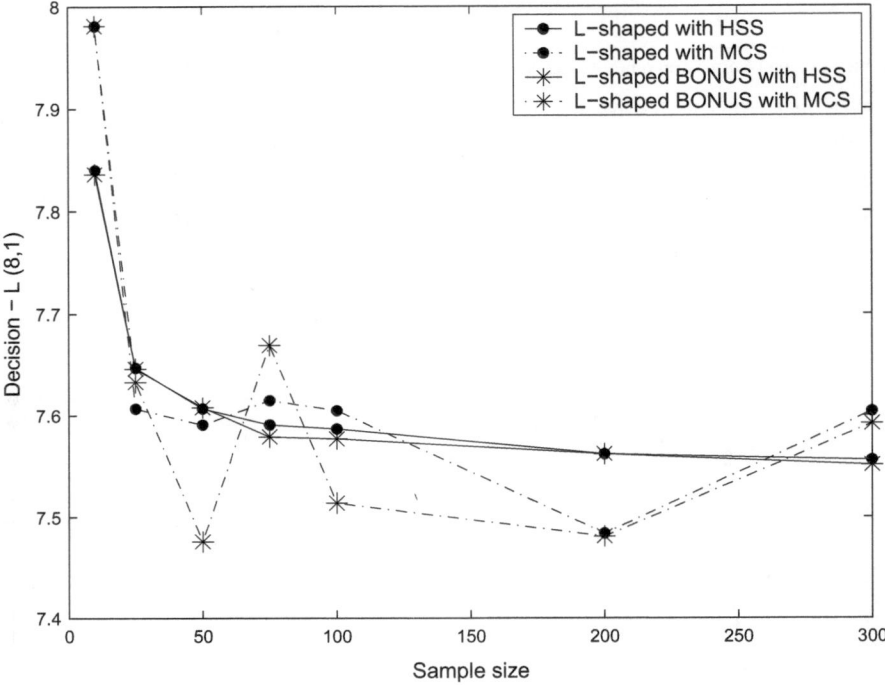

Fig. 9.2 Variation of decision variable L(8,1) with sample size

9.5   Results

To show the efficacy of the L-shaped BONUS method algorithm compared to the standard L-shaped method, and to ensure the benefits of the L-shaped BONUS method are independent of the sampling technique used, the problem was solved in four ways:

- Standard L-shaped method with Hammersley sequence sampling (HSS) technique
- Standard L-shaped method with Monte Carlo sampling (MCS) technique
- L-shaped BONUS method with HSS technique
- L-shaped BONUS method with MCS technique.

For each methodology, the value of the objective function (total cost) is shown as a function of sample size in Fig. 9.1. The value of land allocated by point source 8 toward nitrogen pollution trading ($L(8, 1)$), a representative decision variable, is shown in Fig. 9.2. While it can be seen that the optimization method has a larger effect on the value of the objective function than the sampling technique, the variation among all four methodologies is just larger than 1 %, well within acceptable tolerance limits for the solution of a stochastic nonlinear program. The plot for the decision variable shows an even smaller average difference between the standard L-shaped

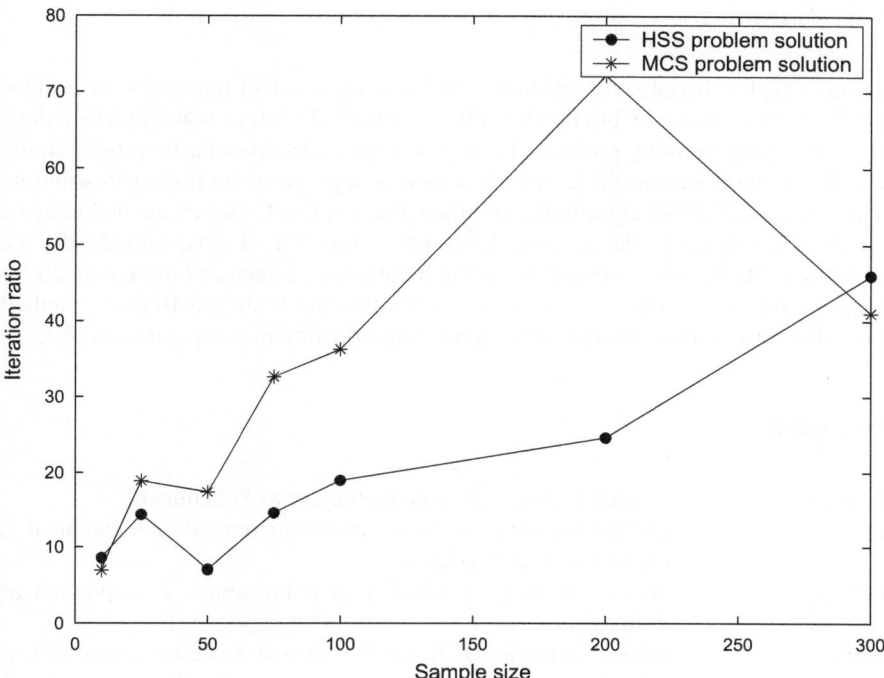

Fig. 9.3 Variation of computational time with sample size

method and the L-shaped BONUS method (0.05 % for HSS and 0.26 % for MCS). This indicates that the reweighting approximation used in L-shaped BONUS method does not sacrifice accuracy in calculating results.

Figure 9.3 shows the iteration ratio, the number of iterations required for the standard L-shaped method divided by the number of iterations required for the L-shaped BONUS method, as a function of sample size. The iteration ratio generally increases as the sample size gets larger; thus, the larger the sample size, the greater the computational savings using the BONUS method. Because Figs. 9.1 and 9.2 show that the value of both the objective function and the representative decision variable reaches a steady-state value as the sample size is increased, indicating that a large sample size is needed for accurate results, it can therefore be concluded that the L-shaped BONUS method is of significant utility in reducing the computational cost of the environmental trading problem.

The solution to the trading problem is qualitatively similar under all four solution methods: to minimize the total cost, every point source needs to achieve part of its required reductions through trading with the nonpoint source. Such a decision is unlikely to result when each point source makes an independent decision, without consideration of the overall cost of reductions. Therefore, the results strongly suggest that a rigorous method and systematic mathematical analysis should be performed to achieve environmental benefits at the lowest possible cost.

9.6 Summary

In this chapter, because the environmental trading problem that seeks to manage nutrient pollution in the Christina River Basin of the LDR is formulated as a two-stage stochastic programming problem, the use of a specialized stochastic programming solution method, such as the L-shaped method, is appropriate for finding its solution. The L-shaped BONUS algorithm, combining the standard L-shaped method with the BONUS reweighting scheme, provides results within 2 % of those provided by the standard L-shaped method while reducing iterations by a factor of more than 20 for large sample sizes. Therefore, it is concluded that the L-shaped BONUS method provides both accurate results and a significant reduction in computational cost.

Notations

$b_{NPS}(j)$	nonpoint source abatement efficiency of chemical j
$c_n(j)$	cost for the nonpoint source discharge reduction of chemical j, expressed as dollars/area
$c_p(i, j)$	cost of treating chemical j at point source i, expressed as dollars/mass
$D(i)$	total volumetric discharge from point source i, expressed as volume/time
$e_{p0}(i, j)$	pretreatment discharge quantity of chemical j from point source i, expressed as mass/volume
e_{n0}	pretreatment discharge quantity of chemical j from the nonpoint source, expressed as mass·area/time
$L(i, j)$	land allocated for trading by point source i to treat chemical j
L_{max}	maximum amount of nonpoint source land available for trading, expressed as area
M	number of regulated chemicals
n	a particular sample from the uncertain space
N_{samp}	the sample size used to represent the uncertain space in the optimization algorithm
P	number of point sources
$q_n(j) = e_{n0}(j) b_{NPS}(j)$	abatement in nonpoint source discharge of chemical j, expressed as mass·area/time
$q_p(i, j)$	discharge abatement of chemical j at PS i (expressed as mass/volume)
R	recourse function
$z_{red}(ij)$	targeted reduction in discharge of chemical j by point source i (expressed as mass/time)
$z_{allowed}(j)$	maximum permitted discharge of chemical j at any single location (expressed as mass/time)
Greek letters	
α_j	cost exponent

Chapter 10
Water Security Networks

10.1 Introduction

Because of the importance of water to all life on Earth, water security has become a critical matter in national and international sustainability. Contamination through either malicious (e.g., terrorist attacks) or accidental means (e.g., industrial accidents) could quickly become a catastrophic event. Therefore, water utilities and their related government agencies perceive a growing need to detect and minimize water contamination in distributed water networks. Much attention has been given to optimization of water network design, such as network capacity, pipe diameter and length, component failures, etc., as a means of minimizing risk (see review for sensor placement in water networks [17]). However, work on qualitative aspects of water network design, such as chemical propagation, concentration of disinfectants, contamination minimization, etc., is far less prominent, though such factors should be considered at the design stage to best make water networks secure from contamination.

One approach given little attention in the literature thus far is the use of sensors in the water distribution network that can detect contaminated water and provide feedback from which appropriate control measures could be taken to minimize the risk. In determining optimal sensor placement, it is necessary to consider the trade-off between the cost of sensors and the risks of contamination. Noise factors and uncertainties in contamination scenarios make it challenging to find a robust solution in the face of risk. One major source of uncertainty is changing water demands at various junctions within the network, which stems from both uncertainty with time (such as the difference in demand between morning and evening) and uncertainty about the actual value at a particular time. Another is varying probability of contamination at a specific node. Therefore, the problem requires a methodology that identifies optimum locations while accounting for these uncertainties, and thus becomes a stochastic programming problem.

In this chapter, the demand uncertainty is extensively modeled, affecting both the constraints and the objective function of the stochastic programming problem. The problem is thereby transformed into a two-stage stochastic programming problem with recourse. The solution is then found using the L-shaped BONUS algorithm. This method gives results that are truly optimal in an actual water distribution network. This chapter is based on Shastri and Diwekar [50].

© Urmila Diwekar, Amy David 2015   127
U. Diwekar, A. David, *BONUS Algorithm for Large Scale Stochastic Nonlinear*
Programming Problems, SpringerBriefs in Optimization, DOI 10.1007/978-1-4939-2282-6_10

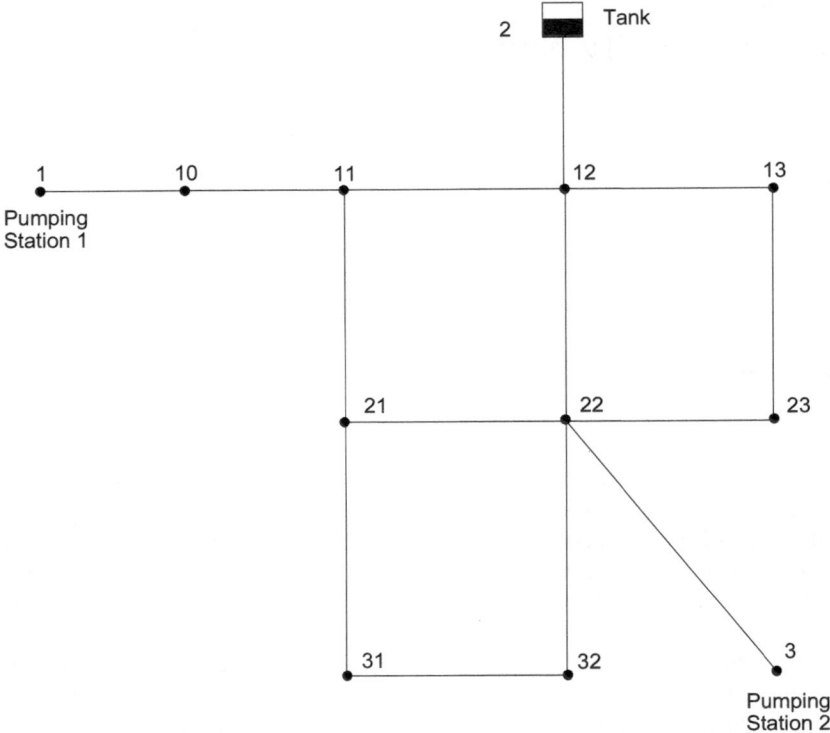

Fig. 10.1 Example water network

10.2   Motivation and Prior Work

A deterministic approach to the water network security problem is explored in [1] .
Here, the risk from contamination is minimized by using sensors for timely detection,
with the objective of minimizing total sensor cost. The problem is formulated as an
integer program (OP), modeling a water network as a graph $G = (V, E)$ (see [3] for
more on graph theory), where E is the set of edges representing pipes and V is the set
of nodes where pipes meet (reservoirs, tanks, consumption points, etc.). An example
network is shown in Fig. 10.1. A contaminating event, or "attack" is modeled as the
release of a large volume of a harmful contaminant at a single point in the network
with a single injection. The water network simulator EPANET is used to determine
the water flow, given the set of available water sources, and assuming that the current
demand pattern holds steady for sufficiently long. The IP formulation is given as

$$\min \quad \Sigma_{i=1}^{n} \Sigma_{p=1}^{P} \Sigma_{j=1}^{n} \alpha_{ip} C_{ipj} \delta_{jp} \tag{10.1}$$

where   (10.2)

$$C_{ipi} = 1, \quad i = 1, ..., n; p = 1, ..., P \tag{10.3}$$

$$s_{ij} = s_{ji}, \quad i = 1, ..., n - 1; i < j \tag{10.4}$$

$$C_{ipj} \geq C_{ipk} - s_{kj} \quad (i, k, j) \in E \, s.t. \, f_{kjp} = 1 \tag{10.5}$$

$$\sum_{(i,j) \in E, i < j} s_{ij} \leq S_{max}; \quad s_{ij} \in \{0, 1\}; (i, j) \in E. \tag{10.6}$$

In this representation, deterministic parameter P is the number of flow patterns and S_{max} is the maximum number of sensors that can be placed. Uncertainty is introduced through α_{ip}, the probability of an attack at node v_i during flow pattern p (conditional on exactly one attack on a node during some flow pattern) and δ_{ip}, the population density at node v_i while flow p is active. Both these sets of parameters are modeled as a known probability distribution (either uniform or normal).

f_{kjp} describes the flow pattern: if water flows from node k to node j in flow pattern p, then $f_{kjp} = 1$ and 0 otherwise. C_{ipj} is the contamination indicator; $C_{ipj} = 1$ if node v_j is contaminated by an attack at node v_i during pattern p, and 0 otherwise. Both f_{kjp} and C_{ipj} are determined through use of the EPANET simulator of water networks (see [40]). Decision variable s_{ij} is 1 if a sensor is placed on the edge (v_i, v_j) and 0 otherwise.

The objective function considers both the probability of an attack and the severity of an attack, the latter influenced by both the number of nodes affected and the population density at each of those nodes, for a given flow pattern. The first set of constraints ensures that a node is contaminated if directly attacked, and the second indicates that a single sensor allows for sensing of water flow in either direction within an edge (pipe). The third set of constraints propagates contamination throughout the water network: if there is positive flow from v_i to v_j and no sensor on that edge, then contamination at v_i results in contamination at v_j. The fourth constraint limits the sensors placed to the total number of sensors available. The final set of constraints enforces the binary nature of the decision variables, which, in turn, enforces the same for the contamination indicators. The IP formulation therefore seeks to minimize the overall impact of an attack, by placing sensors so as to minimize the number of people whose water supply would be contaminated, while remaining within a fixed total for the number of sensors.

However, this formulation is limited in that it assumes that demand at each node is determined solely by the population density at that node. The network flow patterns are therefore determined by the population density. In reality, demand can vary with time of day, season, etc., and thereby change the network flow, rendering the objective function insufficient for capturing the true risk. A similar shortcoming occurs in the constraints, rendering the IP formulation insufficient for producing an optimal result.

In order to better capture the effects of uncertainty, the problem can be modified as follows:

$$\min \quad \Sigma_{l=1}^{N_{samp}} \Sigma_{i=1}^{n} \Sigma_{p=1}^{P} \Sigma_{j=1}^{n} \alpha_{ip}(l) \delta_{jp}(l) C_{ipj} \tag{10.7}$$

where

$$C_{ipi} = 1, \quad i = 1, ..., n; p = 1, ..., P \tag{10.8}$$

$$s_{ij} = s_{ji}, \quad i = 1, ..., n - 1; i < j \tag{10.9}$$

$$C_{ipj} \geq C_{ipk} - s_{kj} \quad (i, k, j) \in E s.t. f_{kjp} = 1 \tag{10.10}$$

$$\sum_{(i,j) \in E, i < j} s_{ij} \leq S_{max}; \quad s_{ij} \in \{0, 1\}; (i, j) \in E. \tag{10.11}$$

The stochastic formulation can be solved using sampling-based techniques, wherein the distributions of the uncertain parameters are approximated by a sample set. Here, N_{samp} is the number of samples to be used in such a technique. This formulation weights the objective function according to the uncertainty of the population density. Further, flow demands are taken to be directly proportional to the population density, so flow throughout the network may change according to the uncertainty of population density. This represents a significant improvement over the initial IP.

However, this formulation fails to consider the cost of sensors. Sensors are available in a broad range of resolutions and accuracy, both of which vary with cost. Considering both the costs of sensors, β_{ij}, and the set of uncertain parameters, u, allows for the formulation of the problem as a two-stage stochastic programming problem with recourse.

First-stage problem:

$$\min \quad \Sigma_{(i,k) \in E} \beta_{ij} s_{ij} + E[R(C, s, u)] \tag{10.12}$$

$$s_{ij} = s_{ji} \quad i = 1, ..., n - 1; i < j \tag{10.13}$$

$$\Sigma_{(i,j) \in E, i < j} s_{ij} \leq S_{max} s_{ij} \in (0, 1); (i, j) \in E \tag{10.14}$$

Second-stage problem:

$$\min \quad E[R(C, s, u)] = \Sigma_{l=1}^{N_{samp}} \Sigma_{i=1}^{n} \Sigma_{p=1}^{P} \Sigma_{j=1}^{n} S\alpha_{ip}(l)\delta_{jp}(l)C_{ipj} \tag{10.15}$$

$$c_{ipi} = 1 \quad i = 1, ..., n; p = 1, ..., P \tag{10.16}$$

$$C_{ipj} \geq C_{ipk} - s_{kj} \quad (i, k, j) \in E; s.t. f_{kjp} = 1 \tag{10.17}$$

Here, the first-stage decision variables are the sensor placement represented by s_{ij}, and the recourse variables are the contamination indicators represented by C_{ipj}. S is the rededication cost associated with each person affected by the contamination (e.g., treatment cost). Note the similarity of the second-stage problem to the modified problem given in Eqs. 10.7 through 10.11; the only difference is the addition of S, so as to properly weight the cost of remediation against the cost of sensors. Again, N_{samp} is used to discretize the uncertain space.

10.3 Solution Methodology

A two-stage stochastic programming problem with recourse, using sampling to approximate continuous uncertain space, can be solved using sampling based optimization methods. Using the L-shaped method, the first-stage problem uses a linear approximation of the second-stage recourse function and the additional constraints sequentially generated in the the second stage to fix the first-stage decision variables. These first-stage decision variables are then used as inputs to the second stage to compute the exact value of the recourse function using the generated scenarios, which is traditionally done by solving the dual of the second-stage problem for every scenario.

In this particular problem, the second-stage problem depends on the values of C_{ipj} and f_{ijp}, thus, the EPANET simulation of the water network would need to be run for every sampled value of δ_{jp}. This would be computationally expensive, as a large number of samples are required, but can be remedied by instead using the L-shaped BONUS algorithm, as described in Chap. 8. The EPANET simulation is therefore performed for only one set of uniform samples. For the subsequent iterations, the BONUS reweighting scheme is used. Note that the Hammersly sequence sampling (HSS) technique is used for sampling because of the advantages provided by its k dimensional uniformity (see [23] and Wang et al. 2004).

10.3.1 Use of BONUS Reweighting for Pattern Estimation

A particular network flow pattern, p, is mathematically defined by the various f_{ijp} values in the network. Use of the BONUS reweighting scheme for determining flow patterns was validated using the following procedure:

1. Take a fixed number of samples from both a uniform distribution and a normal distribution on the set of population densities at each node (δ_{ij}).
2. Perform EPANET simulations for every sample, recording the number of times that binary variable $f_{ijp} = 1$. For example, if 100 simulations are performed, and f_{14p} takes a value of 1 in 56 of the simulations, the value 56 is noted for f_{14p}, indicating that the there was positive water flow from node 1 to node 4 in 56 out of the 100 simulations.
3. Estimate the values for f_{ijp} using the BONUS reweighting scheme,
4. Compare the values found in step 2 for the normal distribution to those calculated in step 3, and calculate an error of estimation as the absolute value of the difference between them.
5. Calculate the standard deviation of the error of estimation.

These steps were repeated for various sample sizes ranging from 100 to 700. As shown in Fig. 10.2, the standard deviation of the estimation error with 700 samples is about 15 % of that with 100 samples. Therefore, the larger sample size provides a significant improvement in the accuracy of the reweighting scheme. Additionally, the number of uncertain nodes was varied from 1 to 8, with the former resulting in an estimation error of 0.13, while the latter resulted in an estimation error of 13.31. Thus, estimation quality degrades with the number of nodes in the network.

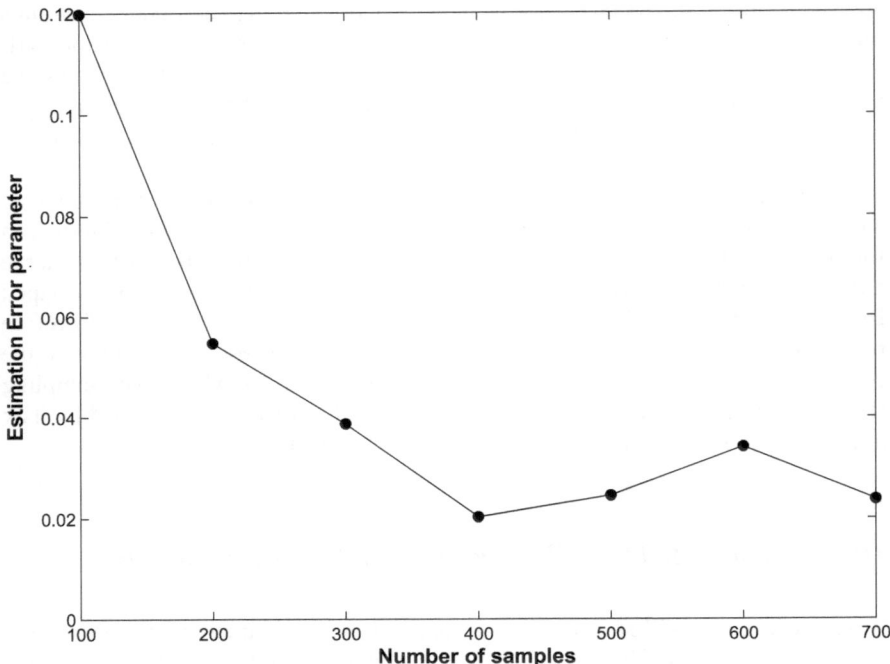

Fig. 10.2 Estimation error as a function of sample size

10.3.2   Back Estimation of Flow Patterns

The BONUS reweighting approach gives the expected values of all f_{ijp} in the given
network for a normal distribution on each δ_{ij}. To solve the sensor placement problem,
these f_{ijp} values must be used to find the frequency of occurrence of various flow
patterns, i.e., how many times a particular flow pattern appears, given a particular
number of simulations, n. This is done through an optimization problem that seeks to
achieve the estimation value of the summations of f_{ij} over all p found in the previous
section, using the f_{ijp} values of various known patterns. For example, if f_{14} is known
to be 56, and the total number of flow patterns to be considered, designated n_p, is
100, then the optimization problem attempts to find a set of 100 flow patterns such
that there is positive flow from node 1 to node 4 in 56 of them.

Mathematically, this problem is represented as

$$\min \Pi_{p=1}^{P} n_p \tag{10.18}$$

$$\text{subject to: } \Sigma_{p=1}^{P} n_p = 100 \tag{10.19}$$

$$\Sigma_{p=1}^{P} n_p f_{ijp} = \Sigma_{m=1}^{N_{samp}} \tilde{f}_{ij}(m) \ \forall i, j, \tag{10.20}$$

where $\tilde{f}_{ij}$ is the total number of times there is positive flow from node i to node j, as estimated by the BONUS reweighting scheme. The result of this optimization problem can then be directly incorporated into the stochastic problem formulation.

10.4 Results

The sensor placement problem was solved for the water network depicted in Fig. 10.1 using three different formulations:

1. Deterministic formulation (method A): Considering the mean demands and two basic flow patterns, i.e., no uncertainty
2. Formulation with noise consideration (method B): Considering uncertain demands only affecting the objective function, i.e., the original formulation modified to include sensor costs and cost per affected person
3. Stochastic formulation (method C): Considering uncertain demands affecting both the objective function and the constraints and solved using the L-shaped BONUS algorithm.

In all three solution methods, the attack probability at various nodes was considered to be fixed and equal during any pattern. Two demand patterns were considered with significant shifts in demand (which may correspond to, for example, differing times of day in an actual network). For methods B and C, the HSS technique was used to generate 100 samples. Additionally, in method C, various flow patterns for the uncertain demands were identified for a base case uniform sample via simulations in EPANET, resulting in ten total flow patterns, eight more than the two basic patterns considered in methods A and B.

Two possible sensor types were considered: a low-cost sensor for $ 1,500,000 per sensor and a high-cost sensor for $ 4,500,000 per sensor. The unit cost of treatment, S, was estimated at $ 30,000. While it is difficult to estimate the cost of treatment as it is dependent on both the exact nature of the contamination and health care costs in a particular location, the $ 30,000 estimate provides a reasonable trade-off between the two types of cost found in the objective function: sensors and treatment.

The problem was solved by varying the maximum number of allowed sensors from 1 to 14. Selected results are presented in Tables 10.1 and 10.2. Table 10.1 gives the value of the objective function and risk output by the solution method used. Table 10.2 gives the results when samples of the uncertain problem parameters are taken and the uncertainty is propagated through the model output by the corresponding solution method. Note that the results for method C are identical in the two tables; this is because method C performs a probabilistic analysis in the initial decision making.

The results show that the actual cost and risk calculated using method C are less than or equal to the actual cost and risk calculated under methods A and B. In the cases where method C provides a lower cost, this indicates that methods A and B result in suboptimal decisions. Though the percentage risk may be higher under method C, the total objective function is still better minimized when using this

Table 10.1 Comparison of estimated cost and risk for different solution methods

Max. # of sensors	Type of sensor	Method A		Method B		Method C	
		Cost ($ x 10^7)	% risk	Cost ($ x 10^7)	% risk	Cost ($ x 10^7)	% risk
1	Low cost	2.1875	29.375	2.1067	29.647	2.2860	32.364
	High cost	2.4875	29.375	2.4045	29.647	2.5860	32.364
2	Low cost	1.8625	22.727	1.7954	22.658	1.9914	25.628
	High cost	2.4625	22.727	2.3937	22.658	2.5860	32.364
4	Low cost	1.7125	16.856	1.6964	18.885	1.8062	18.277
	High cost	2.4625	22.727	2.3937	22.658	2.5860	32.364

Table 10.2 Comparison of actual cost and risk for different solution methods

Max. # of sensors	Type of sensor	Method A		Method B		Method C	
		Cost ($ x 10^7)	% risk	Cost ($ x 10^7)	% risk	Cost ($ x 10^7)	% risk
1	Low cost	2.2860	32.364	2.2860	32.364	2.2860	32.364
	High cost	2.5860	32.364	2.5860	32.364	2.5860	32.364
2	Low cost	1.9914	25.628	2.1193	27.565	1.9914	25.628
	High cost	2.5914	25.628	2.7193	27.565	2.5860	32.364
4	Low cost	1.9738	20.816	1.9106	19.858	1.8062	18.277
	High cost	2.5917	25.628	2.7193	27.565	2.5860	32.364

method. The differences among the estimated objective and risk among methods are as high as 30 % in some cases. Because the estimated values for method C are the actual values, this difference emphasizes the extent of the suboptimality of results found using method A or method B.

The optimal sensor locations shown in Figs. 10.3 and 10.4 are for low and high-cost sensors, respectively. The sensor locations are identified by a notation consisting of the method used and the maximum number of sensors allowed in that particular solution. For example, method B with two low-cost sensors permitted returns optimal locations of sensors along branches 10-11 and 12-22. Therefore, "b2" appears alongside these branches in the diagram.

Branch 10-11 is identified as a sensor location for all methods used. This is unsurprising, given that it is one of the entry points into the network, and water therefore flows from 10 to 11 in the majority of flow patterns studied. However, the placement of additional sensors varies by method. When low-cost sensors are used, methods A and C place a second sensor at branch 3-22, while method B places a sensor at branch 12-12. This is because, in the two basic flow patterns, the flow in branch 12-22 is always positive, thus, water flows from 12 to 22 and method B does not consider a contamination at 3 or 22 to be a risk to node 12. By contrast, method C accounts for the possibility of a reverse flow in this branch, i.e., water may flow

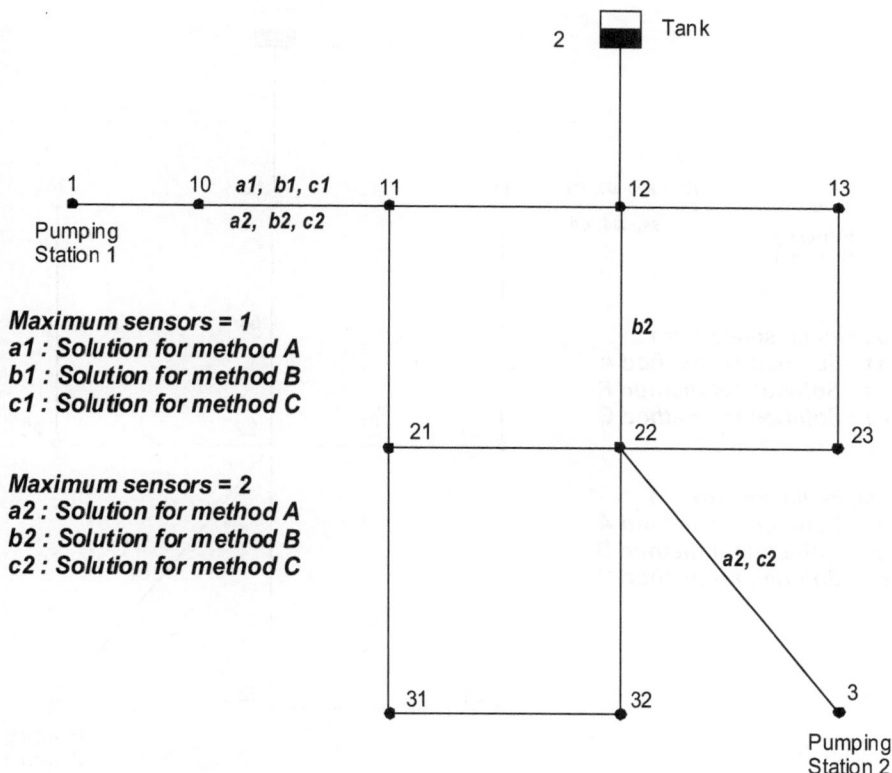

Fig. 10.3 Placement of low-cost sensors for different methods

from 22 to 12, and a contamination at 3 or 22 could indeed affect node 12. Similar reasons can be identified for other differences among solution methods, and we can therefore conclude that method C does a better job of accounting for true risk over methods A and B, which exclude potential real-life events from consideration.

To examine the computational advantages of the L-shaped BONUS algorithm, as compared to the traditional L-shaped algorithm, selected cases were solved using both techniques. For a sample size of 100, use of the L-shaped BONUS method proved to be five times faster. Further, the computational time increases exponentially with the sample size for the standard L-shaped method, but linearly for the L-shaped BONUS method. The comparisons also showed an average difference in the cost and risk percentages of about 4 and 5.4 %, respectively. Thus, accuracy is not compromised through the use of BONUS reweighting.

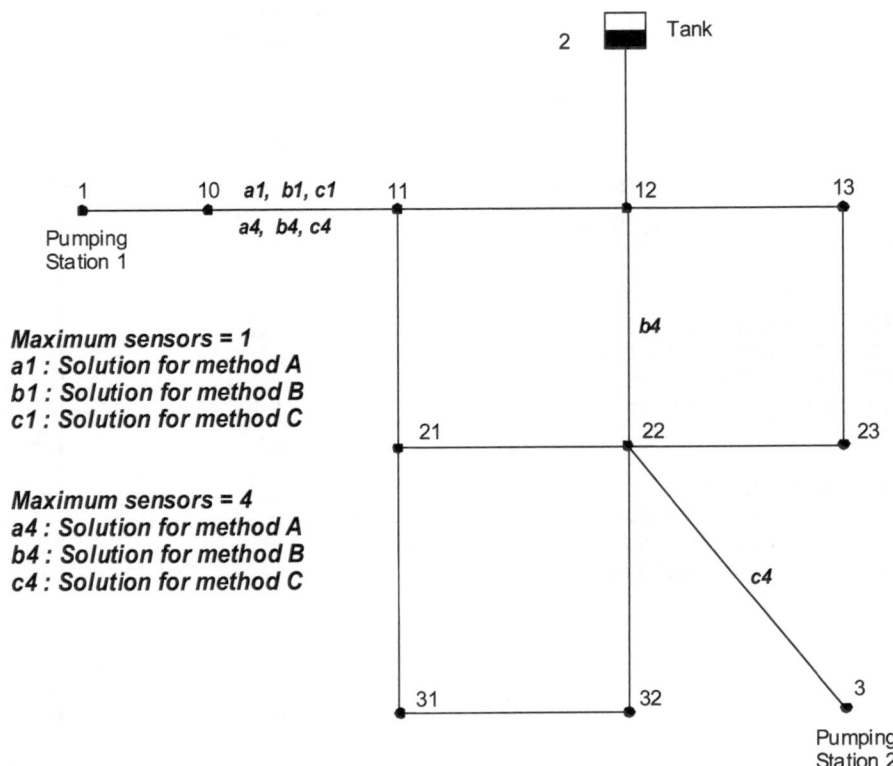

Fig. 10.4 Placement of high-cost sensors for different methods

10.5   Summary

The problem of optimal sensor placement in a water distribution network is critical to securing a water supply, and the trade-offs between risk and the cost of sensing a contamination must be considered in determining optimal sensor locations. Existing problem formulations fail to adequately account for uncertain demands and the multitude of flow patterns that may occur as a result. A stochastic programming formulation that considers uncertainty in both the objective function and the constraints is therefore needed to provide optimal results. The L-shaped BONUS algorithm, which combines the traditional L-shaped method with the BONUS reweighting scheme for flow pattern estimation, allows this formulation to be solved in a short amount of time, and is shown to give accurate results.

Notations

C_{ipj}	contamination indicator; $C_{ipj} = 1$ if node v_j is contaminated by an attack at node v_i during pattern p, and 0 otherwise
E	set of edges representing pipes
f_{ij}	total number of flow patterns that contain positive flow from node v_i to node v_j
f_{kjp}	parameter describing the flow pattern; if water flows from node k to node j in flow pattern p, then $f_{kjp} = 1$ and 0 otherwise
G	graph
n_p	total number of flow patterns to be considered
N_{samp}	number of samples used in a stochastic solution method
P	is the number of flow patterns
S	remediation cost associated with each person affected by the contamination (e.g., treatment cost)
s_{ij}	decision variable representing sensor placement; 1 if a sensor is placed on the edge (v_i, v_j) and 0 otherwise
S_{max}	maximum number of sensors that can be placed
u	set of uncertain parameters
V	set of nodes where pipes meet (reservoirs, tanks, consumption points, etc.)

Greek letters

α_{ip}	the probability of an attack at node v_i during flow pattern p
β_{ij}	cost of a sensor
δ_{ip}	the population density at node v_i while flow p is active

References

1. Berry, J., Fleischer, L., Hart, W., Phillips, C., Watson, J.: Sensor placement in municipal water networks. J. Water Resour. Plan. Manag. **131**(3), 237–243 (2005)
2. Birge, J.R., Louveaux, F.: Introduction to Stochastic Programming. Springer, New York (1997)
3. Bondy, J.A., Murty, U.S.R.: Graph Theory with Applications, vol. 6. Macmillan, London (1976)
4. Charnes, A., Cooper, W.W.: Chance-constrained programming. Manag. Sci. **5**, 73–19 (1959)
5. Dantzig, G.B., Infanger, G.: Large scale stochastic linear programs–importance sampling and Bender decomposition. In: Brezinski, C., Kulisch, U. (eds.) Computational and Applied Mathematics, pp. 111–120 (1991)
6. Dantzig, G.B., Wolfe, P.: The decomposition principle for linear programs. Op. Res. **8**, 101 (1960)
7. Diwekar, U.M.: Introduction to Applied Optimization. Springer, New York (2008)
8. Diwekar, U., Rubin, E.S.: Stochastic modeling of chemical processes. Comput. Chem. Eng. **15**(2), 105 (1991)
9. Diwekar, U.M., Rubin, E.S.: Parameter design method using stochastic optimization with ASPEN. Ind. Eng. Chem. Res. **33**, 292–298 (1994)
10. Evans, L.B., Boston, J.F., Britt, H.I., Gallier, P.W., Gupta, P.K., Joseph, B., Mahlec, V., Ng, E., Seider, W.D., Yagi, H.: ASPEN: An advanced system for process engineering. Comput. Chem. Eng. **3**(1–4), 319 (1979)
11. Fisher, R.A.: On the mathematical foundations of theoretical statistics. Philosophical Transactions of the Royal Society of London. Series A, Containing Papers of a Mathematical or Physical Character. 222, 309–368 (1922)
12. Frieden, B.R., Gatenby R.A. (eds.): Exploratory data analysis using Fisher information. Springer, London (2010)
13. Gray, F.: Petroleum Production for the Nontechnical Person. PennWell Books, Tulsa (1986)
14. Halton, J.H.: On the efficiency of certain quasi-random sequences of points in evaluating multi-dimensional integrals. Numer. Mathematik **2**(1), 84–90 (1960)
15. Hamilton, T.H.: Power Engineering 81(3), 52 (1977)
16. Hammersley, J.M.: Monte Carlo methods for solving multivariate problems. Ann. New-York Acad. Sci. **86**, 844 (1960)
17. Hart, W., Murray, R.: Review of sensor placement strategies for contamination warning systems in drinking water dstribution systems. J. Water Res. Plan. Manag. **136**, 611 (2010)
18. Hesterberg, T.: Weighted average importance sampling defensive mixture distributions. Technometrics **37**, 185 (1995)
19. Hesterberg, T.: Estimate and confidence intervals for importance sampling sensitivity analysis. Mathl. Comput. Model. **23**(8/9), 79 (1996)
20. Higle, J.L., Sen, S.: Stochastic decomposition: an algorithm for two-stage linear programs with recourse. Math. Oper. Res. **16**(3), 650–669 (1991)

© Urmila Diwekar, Amy David 2015
U. Diwekar, A. David, *BONUS Algorithm for Large Scale Stochastic Nonlinear Programming Problems,* SpringerBriefs in Optimization, DOI 10.1007/978-1-4939-2282-6

21. Iman, R.L., Conover, W.J.: Small sample sensitivity analysis techniques for computer models, with an application to risk assessment. Commun. Stat. **A17**, 1749 (1982)
22. James, B.A.P.: Variance reduction techniques. J. Oper. Res. Soc. **36**(6), 525–530 (1985)
23. Kalagnanam, J.R., Diwekar, U.M.: An efficient sampling technique for off-line quality control. Technometrics **39**, 308 (1997)
24. Kauffman, G.J., Wozniak, S.L., Vonck, K.J.: Executive Summary: A Watershed Restoration Action Strategy (WRAS) for the Delaware Portion of the Christina Basin, University of Delaware, College of Human Services, Education, and Public Policy, Institute for Public Administration, Water Resources Agency (2003)
25. Kocis, L., Whiten, W.: Computational investigations of low discrepancy sequences. ACM Trans. Math. Softw. **23**(2), 266–294 (1997)
26. Knuth, D.E.: The Art of Computer Programming, volume 1: Fundamental Algorithms. Addison-Wesley, US (1937)
27. Ku, A.: Modelling uncertainty in electricity capacity planning, Ph.D. Thesis University of London-London Business School, 445 (1995)
28. Lee, A.J., Diwekar, U.M.: Optimal sensor placement in integrated gasification combined cycle power systems. Appl. Energy **99**, 255 (2012)
29. Lehmer, D.H.: Mathematical methods in large scale computing units. Proceedings of the 2nd Symposium on Large Scale Digital Calculating Machinery, 141 (1949)
30. Louveaux, F., Smeers, Y.: Optimal Investments for Electricity Generation: A Stochastic Model and Test-Problem. Springer Series in Computational Mathematics, 445. Springer, Berlin (1988)
31. McKay, M.D., Beckman, R.J., Conover, W.J.: A Comparison of three methods of selecting values of input variables in the analysis of output from a computer code. Technometrics **21**(2), 239 (1979)
32. Metropolis, N., Ulam, S.: The Monte Carlo method. J. Am. Stat. Assoc. **44**(247), 335 (1949)
33. Morgan, G., Henrion, M.: Uncertainty: A Guide to Dealing with Uncertainty in Quantitative Risk and Policy Analysis. Cambridge University Press, Cambridge (1990)
34. Niederreiter, H.: Random Number Generation and Quasi-Monte Carlo Methods. SIAM Publications, Philadelphia (1992)
35. NETL: EPA Study-Case 8: IGCC, Technical Report DOE/NETL-401/042606, U.S. Department of Energy National Energy Technology Laboratory, November (2006)
36. NETL: Cost and Performance Baseline for Fossil Energy Plants. Volume 1: Bituminous Coal and Natural Gas to Electricity, Technical Report DOE/NETL-2010/1397, National Energy Technology Laboratory (2010)
37. Rhodes, A.: Refinery operating variables key to enhanced lube oil quality. Oil Gas J. **91**(1), 45–51 (1993)
38. Rockafellar, R.T., Wets, R.J.-B.: Stochastic convex programming: Basic duality. Pacific J. Math. **63**, 173 (1976)
39. Rockafellar, R.T., Wets, R.J.-B.: Scenarios and policy aggregation in ptimization under uncertainty. Math. Oper. Res. **16**, 119 (1991)
40. Rossman, L.A.: EPANET 2: Users manual (2000)
41. Rubin, E.S., Kalagnanam, J.R., Frey, H.C., Berkenpas, M.B.: Integrated environmental control modeling (IECM) of coal-fired power systems. J. Air Waste Manag. Assn. **47**, 1180 (1997)
42. Ruszczynski, A.: A regularized decompotision for minimizing a sum of polyhedral functions. Math. Program. **35**, 309 (1986)
43. Sahin, K., Diwekar, U.: BONUS: Better optimization algorithm for nonlinear uncertain systems. Ann. Oper. Res. **132**, 47 (2004)
44. Salazar, J.M., Diwekar, U.M.: An efficient stochastic optimization framework for studying the impact of seasonal variation on the minimum water consumption of pulverized coal (PC) power plants. Energy Syst. **2**(3–4), 263–279 (2011)
45. Salazar, J.M., Zitney, S., Diwekar, U.M.: Minimization of water consumption under uncertainty for a pulverized coal power plant. Environ. Sci. Technol. **45**(10), 4645–4651 (2011)

46. Salazar, J., Diwekar, U., Constantinescu, E., Zavala, V.: Stochastic optimization approach to water management in cooling-constrained power plants. Appl. Energy **112**, 12–22 (2013)

47. Saliby, E.: Descriptive sampling: a better approach to Monte Carlo simulations. J. Oper. Res. Soc. **41**(12), 1133 (1990)

48. Scatena, F., et al.: Water quality trading in lower delaware river basin: a resource for practitioners. Technical report, Institute for Environmental Studies, University of Pennsylvania, Philadelphia, PA (2006)

49. Shastri, Y., Diwekar, U.: An efficient algorithm for large scale stochastic nonlinear programming problems. Comput. Chem. Eng. **30**, 864 (2006a)

50. Shastri, Y., Diwekar, U.: Sensor placement in water networks: a stochastic programming approach. J. Water Res. Plan. Manag **132**, 192 (2006b)

51. Shastri, Y.N., Diwekar, U..M.: L-Shaped BONUS algorithm with application to water pollutant trading. Ind. Eng. Chem. Res. **47**, 9417 (2008)

52. Silverman, B.W.: Density Estimation for Statistics and Data Analysis. Chapman & Hall (CRC reprint 1998), Boca Raton (1986)

53. Skamarock, W.C., Klemp, J.B., Dudhia, J., Gill, D.O., Barker, D.M., Duda, M.G., Huang, X.-Y., Wang, W., Powers, J.G.: A Description of the Advanced Research WRF Version 3. Technical Report Tech Notes-475+ STR, NCAR (2008)

54. Subramanyan, K., Diwekar, U.M.: User Manual for Fortran-Based Stochastic Sampling Code, Center for Uncertain Systems: Tools for Optimization and Management (CUSTOM). Vishwamitra Research Institute, Clarendon Hills (2006)

55. Taguchi, G.: Introduction to Quality Engineering. Asian Productivity Center, Tokyo (1986)

56. Tietenberg, T.H.: Emissions trading, an exercise in reforming pollution policy. Resources for the Future (1985)

57. US EPA: Water Quality Trading Policy, Technical Report, United States Environmental Protection Agency (USEPA), Office of Water, Washington, DC (2003)

58. Van Slyke, R., Wets, R.J.B.: L-shaped linear programs with application to optimal control and stochastic programming. SIAM J. Appl. Math. **17**, 638 (1969)

59. Wang, R., Diwekar, U., Padro, C.E.G.: Efficient sampling techniques for uncertainties in risk analysis. Environ. Prog. **23**, 141 (2005)

60. Wichmann, B., Hill, I.: Algorithm AS 183. An efficient and portable pseudo-random number generator. Appl. Stat. **31**, 188 (1982)

Index